T0258165

Applied Principles of Nanofabrication

Applied Principles of Nanofabrication

Edited by **Lindy Bowman**

New York

Published by NY Research Press,
23 West, 55th Street, Suite 816,
New York, NY 10019, USA
www.nyresearchpress.com

Applied Principles of Nanofabrication
Edited by Lindy Bowman

© 2015 NY Research Press

International Standard Book Number: 978-1-63238-056-2 (Hardback)

Printed in the United States of America.

Contents

Preface

Nanofabrication is an active field of research. Nanotechnology has witnessed a rapid expansion in the past few years, mainly owing to the speedy development in nanofabrication methods employed to manufacture nano-devices. Nanofabrication can be divided into two portions: methods using chemical combination and methods using nanolithography. This book contains various chapters and aims to offer the essential and current developments of nanolithography. It discusses significant issues regarding lithographic technologies such as EUV lithography, scanning probe, nano-sphere and inkjet printing. The book also covers various aspects related to resolution enhancement techniques. Most chapters analyze the lithographic procedures available for researchers and experts. This book covers the subject thoroughly and will be beneficial for its readers.

The information contained in this book is the result of intensive hard work done by researchers in this field. All due efforts have been made to make this book serve as a complete guiding source for students and researchers. The topics in this book have been comprehensively explained to help readers understand the growing trends in the field.

I would like to thank the entire group of writers who made sincere efforts in this book and my family who supported me in my efforts of working on this book. I take this opportunity to thank all those who have been a guiding force throughout my life.

Editor

Part 1

EUV Lithography and Resolution Enhancement Techniques

High-Index Immersion Lithography

Keita Sakai
Canon Inc.
Japan

1. Introduction

The resolution capability of photolithography is given by Rayleigh's equation.

$$R=k_1 \cdot \lambda/NA , \qquad (1)$$

where R is the half-pitch resolution of the image, k_1 is a constant that depends on the resist process and exposure method, λ is the exposure wavelength, and NA is the numerical aperture of the projection optic. According to Rayleigh's equation, there are three ways to enhance the resolution. The first is to shorten the exposure wavelength such as extreme ultraviolet lithography (EUVL). The second is to improve the k_1 value, for example, using the double-patterning technique. The third is to increase the numerical aperture (NA) as ArF immersion lithography. It has already realized the NA up to 1.35 and moreover can increase the NA using high-index materials. In this chapter, high-index immersion lithography with the NA over 1.45 is focused.

The NA is actually determined by the acceptance angle of the lens and the refractive index of the medium surrounding the lens and is given by eq. (2).

$$NA=n \cdot \sin\theta , \qquad (2)$$

where n is the refractive index of the medium surrounding the lens and θ is the acceptance angle of the lens. Therefore, the NA can be enlarged by replacing the air (n=1) with a fluid (n>1) as a medium. In this immersion lithography, the film stack consists of a lens, a fluid layer, and a resist layer. The value of $n \cdot \sin\theta$ is invariant through the film stack because it obeys Snell's Law. Since $\sin\theta$ is smaller than 1, the maximum NA (=$n \cdot \sin\theta$) is limited by the layer with the minimum refractive index. For example, the refractive index of water is 1.44 at 193.4 nm, thus the NA over 1.44 cannot be established as shown in Fig. 1 (a) because of the total reflection. To realize the NA over 1.44, the water must be replaced with a fluid which has a higher refractive index than water (Fig. 1 (b)). In Fig. 1 (b), fused silica has the smallest refractive index in the film stack and it limits the maximum NA. For further increasing the NA, a high-index lens material must be used as a lens material.

As described above, high-index lens materials and high-index immersion fluids are indispensable to realize high-index immersion lithography. One of the candidates of a high-index lens material is lutetium aluminium garnet (LuAG), which has a refractive index of 2.14. Second-generation (G2) and third-generation (G3) fluids are saturated hydrocarbon fluids whose refractive indices are approximately 1.64 and 1.80, respectively.

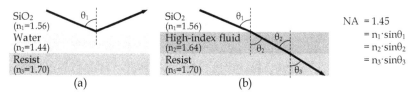

Fig. 1. Light propagation through the film stack at 1.45 NA. (a) The light does not reach the resist layer because of the total reflection at the lens-water boundary. (b) A high-index fluid enables the light to propagate into the resist.

High-index immersion lithography can be classified into three types by combining a lens material and an immersion fluid: fused silica and a G2 fluid (type 1), LuAG and a G2 fluid (type 2), and LuAG and a G3 fluid (type 3). With these types, the maximum NAs are estimated 1.45, 1.55, and 1.70, respectively.

Table 1 shows k_1 values for typical half-pitch and NA. The k_1 is calculated using Rayleigh's equation and needs to be at least 0.25 to resolve the patterns of the half-pitch for the theoretical limit. The resolution of 36 nm is achieved with 1.45 NA optic, which can be realized using only a G2 fluid. Those of 1.55 NA and 1.65 NA can achieve the resolutions of 34 and 32 nm, respectively.

Half-pitch	1.35 NA	1.45 NA	1.55 NA	1.65 NA	1.70 NA
39 nm	0.272	0.292	0.313	0.333	0.343
36 nm	0.251	0.270	0.289	0.307	0.316
34 nm	0.237	0.255	0.272	0.290	0.299
32 nm	0.223	0.240	0.257	0.273	0.281

Table 1. k_1 values for typical half-pitch and NA.

Although an exposure tool of 1.45 NA does not need new materials except for a G2 fluid, customers have little interest in the tool because of its modest gain in resolution. On the other hand, a tool of over 1.65 NA seems attractive for resolution enhancement. However, it would be difficult to realize G3 fluids immediately because no materials meet the requirements. In such a situation, it is a realistic way to develop a tool of 1.55 NA using LuAG and a G2 fluid.

In the next subchapter, projection optics with LuAG are explained. Development of LuAG is a hard work because the specifications of lithography-grade lens materials are extremely stringent. According to the history of LuAG development by Schott Lithotec (Parthier et al., 2008), great progress was achieved in the absorbance but it does not reach the target. The key issue for the optical design with LuAG is a correction of intrinsic birefringence (IBR). An effective method has been developed for IBR correction and it reduces the wave-front aberration to a practical level.

In the third subchapter, an immersion system using a G2 fluid is described. It was demonstrated that fluid absorbance can be kept low enough through an in-line purification unit and an oxygen removal unit. Although lens contamination is an important issue in a G2 fluid system, it was found that contamination can be suppressed by addition of a small amount of water into a G2 fluid. With this water-addition and in-line purification, the necessity for lens cleaning decreases from three times per day to once a week. Fluid confinement is also a challenge for a G2 fluid system because residual fluid is easy to remain on a wafer. Some issues arising with residual fluid, such as fluid darkening due to reentry of oxygen-rich residual fluid, were solved. By accepting residual fluid on a wafer, the scanning speed and the throughput can be raised.

Finally, the remaining challenges to realize high-index immersion lithography are discussed.

2. Projection optics with LuAG

Key parameters for high-index lens materials are refractive index, absorbance, and intrinsic birefringence (IBR). High-index lens materials must have a refractive index to permit NA scaling sufficient to justify the development cost. The absorbance must be sufficiently low to avoid the image degradation by thermal aberrations. The intrinsic birefringence must be minimal to allow a correction to avoid introducing unacceptable aberrations in the final aerial image.

The National Institute of Standards and Technology (NIST) has searched for high-index lens materials that meet the above requirements such as barium lithium fluoride (BaLiF₃) and LuAG (Burnett et al., 2006). BaLiF₃ developed by Tokuyama is available in various sizes with low absorbance (Nawata et al., 2007). However, the refractive index of BaLiF₃ is not high enough to achieve a sufficient NA for the enhancement of the resolution. Only LuAG remains as a candidate for a high-index lens material. LuAG has the intrinsic birefringence over 30 nm/cm and still has a high absorbance caused by impurities. Thus, the status of LuAG should be paid attention to and an IBR correction method should be developed.

2.1 Status of LuAG

Lithography-grade LuAG has been aggressively developed by Schott Lithotec since 2005. The absorbance, which is the biggest issue for LuAG, was largely improved down to 0.035 /cm by purifying the raw material and optimising the crystal growing process as shown in Fig. 2. Since it was found that the intrinsic absorbance of LuAG is 0.00118 /cm (Letz et al., 2010), the absorbance will be less than 0.005 /cm by further reduction of impurities.

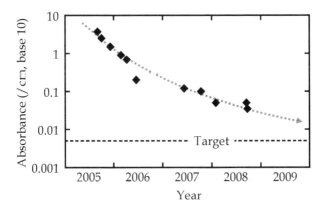

Fig. 2. History of LuAG development by Schott Lithotec. The absorbance of LuAG has been improved down to 0.035 /cm but it has not reached the target (0.005 /cm).

Table 2 shows the target and the status of each requirement. The stress birefringence (SBR) and the homogeneity seem to have gotten closer to the targets. However, these were achieved with a small crystal of yttrium aluminium garnet (YAG). Therefore, it is necessary to confirm the SBR and the homogeneity of large-sized LuAG.

The remaining challenges, such as the absorbance, will be solved if the development is accelerated by strong supports from the industry.

	Target	Status
Absorbance (/cm, base 10)	< 0.005	0.035
Stress birefringence (RMS, nm/cm)	< 0.5	0.73 (40 mmφ, YAG)
Index homogeneity (-Z36, ppm)	< 0.05	0.03 (45 mmφ, YAG)

Table 2. Target and status of LuAG.

2.2 Intrinsic birefringence correction

Since the absorbance of LuAG is higher than that of fused silica, it is difficult to use many LuAG lenses in a projection optic. From the viewpoint of optical design, applying LuAG to the final lens is most effective. In the final lens through which light converging at an image plane passes, the range of incident angles to the optical axis is wide. Assuming that the bottom surface of the final lens, which touches an immersion fluid, is flat, the maximum ray angles in LuAG are 46.4 and 52.6° for the NAs of 1.55 and 1.70, respectively.

The IBR distribution depends on the direction of a crystal. The distribution of a (111)-oriented crystal has three-fold rotational symmetry. When the (111)-oriented crystal is used, the angle at the maximum IBR is 35.3° from the (111) axis. On the other hand, a (100)-oriented crystal has four-fold rotational symmetry, and the angle at the maximum IBR is 45° from the (100) axis. In short, the maximum ray angle exceeds both angles of the maximum IBR even in the case of NA 1.55.

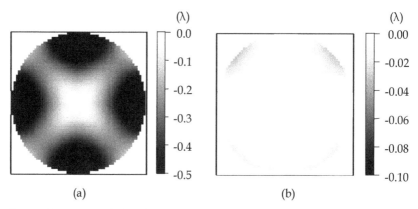

Fig. 3. Difference between radial and tangential polarized wave front for a projection optic of 1.55 NA; (a) without IBR correction, (b) with an original correction method.

Figure 3 shows wave front aberration maps for a projection optic of 1.55 NA. These show the difference between the radial and the tangential polarized wave fronts. The wave front difference without the correction has a large rms value of 217 mλ and a clear 4θ component. On the other hand, the original correction method reduces the rms difference to 4.2 mλ. This value is within the practical level.

In the case of 1.70 NA, the wave front without the correction has a large rms difference of 680 mλ and the rms difference after the correction is 21.2 mλ. Although this value is still large compared with the case of 1.55 NA, it is close to the practical level. Therefore, it can be concluded that the IBR correction is feasible even at 1.70 NA.

3. Immersion system using a G2 fluid

High-index immersion fluids are categorized as second-generation (G2) or third-generation (G3) fluids. G2 fluids have refractive indices of approximately 1.64 and enable an NA to increase up to 1.55 with LuAG as a final lens material. G3 fluids are targeted to have refractive indices over 1.80 for achieving an NA over 1.65 with LuAG.

DuPont, JSR, and Mitsui Chemicals have reported both G2 and G3 fluids (French et al, 2007, Furukawa et al., 2007, Kagayama et al., 2007). Their G2 fluids are saturated hydrocarbon and have sufficient performance as an immersion fluid. They also developed G3 fluids as some extension of organic G2 fluid materials. This type of G3 fluids has not yet met the requirements of refractive index or absorbance. The other types of G3 fluids, which include nanoparticles, are being developed, but they are still within the research phase (Zimmerman et al., 2008). In the above situation, an immersion system using a G2 fluid has been preferentially developed (Sakai et al., 2008).

To achieve the target cost of $1/layer for a G2 fluid, the recycling of a G2 fluid is required. G2 fluids are easily degraded by laser irradiation and the dissolution of oxygen from the atmosphere. Therefore, a fluid circulation system is desired to keep the fluid absorbance sufficiently low using purification and oxygen removal functions. In addition, lens contamination, bubble, and residual fluid on a wafer are also important issues. An immersion system composed of a fluid circulation system, an immersion nozzle and a cleaning unit should be established to accommodate a G2 fluid.

3.1 Fluid degradation

The absorbance of water does not change largely with dissolved oxygen, and immersion water is not used repeatedly as a G2 fluid. These are the reasons the degradation of water is not a concern as an immersion fluid.

On the other hand, the absorbance of a G2 fluid is easily increased by dissolved oxygen or laser irradiation. Moreover, the dn/dT of a G2 fluid is 5~6 times larger than that of water (French et al., 2006, Santillan et al., 2006, Furukawa et al., 2007). This means that a temperature change causes a larger change in the refractive index and results in larger thermal aberration (Sekine et al., 2007). To suppress a temperature rise with the absorption of exposure light, the fluid absorbance has to remain sufficiently low.

3.1.1 Dissolved oxygen

Although the absorbance of a G2 fluid is lower than that of water (French et al., 2005, Wang et al., 2006), the degradation of a G2 fluid with dissolved oxygen is larger than that of water. The first reason for this large degradation is the high oxygen solubility. It has been reported that the solubility of oxygen in G2 fluids are 55~70 ppm in air atmosphere (Furukawa et al., 2006). These values are 7~9 times higher than that in water. According to the research at Columbia University (Gejo et al., 2007), dissolved oxygen forms a charge-transfer complex with cycloalkane of a G2 fluid and the complex has strong absorption in the ultraviolet wavelength range. This is the second reason for the large degradation induced by dissolved oxygen.

The induced absorbance against dissolved oxygen in a G2 fluid has been experimentally obtained. Figure 4 shows the induced absorbance of HIL-203 (JSR) compared with water. It was confirmed that the oxygen-rich HIL-203 exhibits strong absorption. Therefore, a removal function of oxygen is necessary to keep the fluid absorbance low in a circulation system.

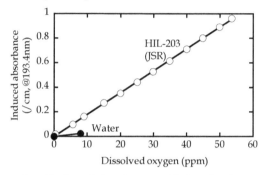

Fig. 4. Experimental results of induced absorbance with dissolved oxygen. Open circles show the induced absorbance of HIL-203 and filled circles show that of water.

An immersion nozzle supplies a G2 fluid under a final lens and sucks the fluid with the neighbouring gas. If the atmosphere around a wafer is air, the oxygen concentration in the recovered fluid increases with the sucked air. Under such a condition, two types of oxygen removal functions were evaluated using an experimental system as shown in Fig. 5. The fluid used in this experiment was HIL-203 (JSR) and the fluid flow rate was 400 mL/min.

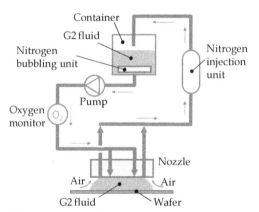

Fig. 5. Schematic view of an experimental system to investigate two types of oxygen removal functions. One is a nitrogen bubbling unit in the fluid container, and the other is a nitrogen injection unit attached to the fluid line. Oxygen concentration in the fluid is measured at the position just before the immersion nozzle.

Figure 6 shows dissolved oxygen using two types of oxygen removal functions. In the case with the nitrogen bubbling unit alone, the oxygen concentration was 4.5 ppm. It corresponds to the absorption of 1.83 %/mm, which can not be permitted. By using the nitrogen bubbling and the nitrogen injection units simultaneously, the oxygen concentration was reduced to 0.2 ppm. Further reduction was achieved increasing the nitrogen flow rate into the injection unit from 5 to 10 L/min. The oxygen concentration of 0.1 ppm is the permissible level because it corresponds to only 0.04 %/mm degradation. The nitrogen injection unit removes the dissolved oxygen from the oxygen-rich fluid just after recovery and prevents the oxygen-rich fluid from going back to the container. That is why the injection unit is efficient for the reduction of dissolved oxygen.

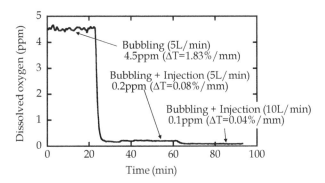

Fig. 6. Experimental results of dissolved oxygen using oxygen removal functions. At first, the dissolved oxygen was not sufficiently low with the nitrogen bubbling unit alone. Then, the large reduction of dissolved oxygen was achieved with the simultaneous use of the nitrogen bubbling and the nitrogen injection units.

3.1.2 Laser irradiation

The absorbance of a G2 fluid increases with the photodecomposed materials of a G2 fluid induced by ArF laser irradiation. To achieve the target cost of $1/layer, it is necessary to remove the photodecomposed materials using an in-line purification unit.

Figure 7 shows the instruments used for a laser irradiation test. ArF excimer laser G41A3 (Gigaphoton), which can irradiate with the rep. rate of up to 3 kHz, is used for irradiation. The irradiation chamber has two beam lines and is purged with nitrogen gas to keep the oxygen concentration less than 1 ppm. The fluid circulation system with the oxygen removal functions can maintain dissolved oxygen at 0.1 ppm and below. The fluid absorbance monitor contains four flow-through cells which can evaluate a number of fluids simultaneously.

The experimental conditions are shown in Table 3. Fluid suppliers have developed in-line purification units for their G2 fluid. Therefore, fluid degradation was evaluated with the combination of a fluid and a purification unit made by the same supplier, such as HIF-A and Unit-A.

The experimental results are shown as the induced absorbance against fluid dose. The fluid dose is defined as an incident dose modified by the volume dilution factor as shown in eq. (3). It has been suggested as the appropriate dose metrics (Liberman et al., 2007).

$$D_{fluid}=I \cdot N \cdot V_{irr}/V_{total} , \qquad\qquad (3)$$

where D_{fluid} is the fluid dose, I is the fluence per pulse, N is the total pulse count, and V_{irr} and V_{total} are the irradiated volume and the total fluid reservoir volume, respectively.

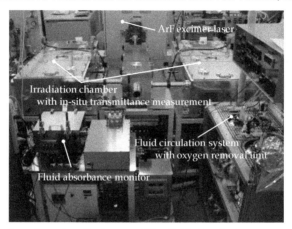

Fig. 7. Instruments for a laser irradiation test. The fluid circulation system consists of a fluid container, a pump, and oxygen removal units.

Immersion fluid / Purification unit	HIF-A / Unit-A HIF-B / Unit-B HIF-C / Unit-C
Fluence	0.5 mJ/(cm² · pulse)
Rep. rate	2000 Hz
Irradiation area	8.0 mmϕ
Fluid gap	1.0 mm
Fluid flow rate	100 ~ 350 mL/min

Table 3. Experimental conditions for evaluation of in-line purification units.

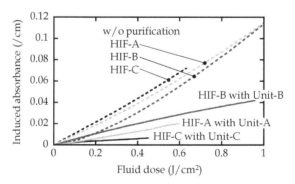

Fig. 8. Induced absorbance of a G2 fluid against the fluid dose. The dashed line shows the induced absorbance without a purification unit. The solid line shows the result with a purification unit. The fluid dose is calculated using eq. (3) and the conditions as shown in Table 3.

Figure 8 shows the induced absorbance of each G2 fluid with and without an in-line purification unit. For example, the degradation rate of HIF-C with Unit-C is one-seventh-fold smaller than that without Unit-C. Using a purification unit, fluid lifetime will be up to 1 week before the fluid absorbance will reach an unacceptable level. Therefore, it can be concluded that recycling of a G2 fluid is feasible using an in-line purification unit.

3.2 Lens contamination

It has been reported that the exposed surface of a final lens is polluted by photodecomposed materials of a G2 fluid (French et al., 2007). This lens contamination diminishes the uniformity of exposure dose on a wafer and would be a source of particles. Thus, the suppression of lens contamination is a serious issue for an immersion system using a G2 fluid.

The suppression of lens contamination was examined by three procedures. The first is to select an appropriate fluid, which has a lower deposition rate. The second is to use an in-line purification, which can remove photodecomposed materials. The third is water-addition into a G2 fluid, which is an original suppression method.

Figure 9 shows a schematic diagram of the experimental setup. The cell absorbance, which consists of the absorbance of windows and a fluid of 1 mm path length in the cell, was measured using two power meters. In addition, the fluid absorbance was measured by a fluid absorbance monitor. Then, the absorbance of windows was calculated by subtracting the fluid absorbance from the cell absorbance as shown in eq. (4).

$$A_{window} = A_{cell} - A_{fluid} , \tag{4}$$

where A_{window} is the absorbance of windows, A_{cell} is the cell absorbance, and A_{fluid} is the fluid absorbance through 1 mm path length.

The laser fluence was 1.0 mJ/(cm²·pulse), which is almost the same as in an actual exposure tool. The dose estimation on an actual tool is approximately 60 kJ/(cm²·day). Although the effective cleaning method was proposed (Liberman et al., 2007), the necessity for lens cleaning should be less than once a week for minimizing tool downtime.

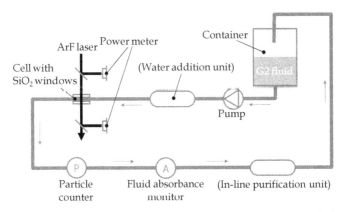

Fig. 9. Schematic diagram of the experimental setup. G2 fluid flows through the 1 mm gap space between the SiO₂ windows of the irradiation cell. The particle counter can detect particles larger than 60 nm in diameter.

3.2.1 Appropriate fluid

Induced absorbance of windows, which correspond to a final lens of a projection optic, was evaluated by using various G2 fluids. Figure 10 shows induced absorbance of windows against exposure dose. HIF-C exhibits a higher deposition rate than HIF-A and D. Therefore, HIF-A or D should be selected from the point of view of lens contamination. However, lens contamination occurs in a few hours on an actual exposure tool, even if using HIF-A or D. It is necessary to introduce some methods for suppressing contamination.

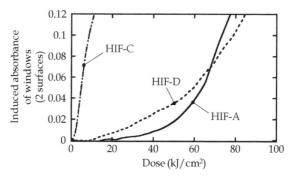

Fig. 10. Induced absorbance of windows using various G2 fluids. The deposition rates of HIF-A (solid line) and HIF-D (dashed line) are lower than that of HIF-C (dashed-dotted line).

3.2.2 In-line purification

It is expected that an in-line purification unit can suppress lens contamination because it can reduce photodecomposed materials. The experiments with several G2 fluids and in-line purification units were carried out using the experimental setup as shown in Fig. 9.

Figure 11 shows experimental results of window absorbance with and without a purification unit. It was confirmed that each purification unit can improve lens contamination. But the performance of purification units are not enough and a final lens needs to be cleaned more than once a day under the practical use conditions. Therefore, the additional method is desired to extend the cleaning interval.

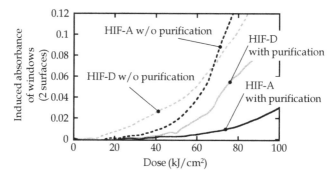

Fig. 11. Induced absorbance of windows with and without a purification unit. The deposition rate with a purification unit (solid line) is lower than that without a purification unit (dashed line).

3.2.3 Water-addition

It was found that lens contamination can be suppressed by addition of a small amount of water into a G2 fluid (Sakai et al., 2008). Since water in a G2 fluid is removed by an oxygen removal unit or an in-line purification unit, a water-addition unit is necessary for a fluid circulation system to keep water in a G2 fluid. Figure 12 shows a schematic diagram of a water-addition unit. Using the water-addition unit, a sufficient amount of water can be solved into a G2 fluid.

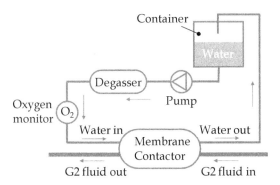

Fig. 12. Schematic diagram of a water-addition unit. The unit consists of a membrane contactor and a circulation unit of degassed water. Degassed water contacts a G2 fluid through the porous membrane with a large area.

Figure 13, 14, and 15 shows the experimental results using HIL-203 (JSR) as a G2 fluid. In the case without water-addition, the absorbance of windows increased gradually (Fig. 13) and many streak contaminations were observed on the window (Fig. 14 (a)). On the other hand, the window was not contaminated in the case with water-addition (Fig. 13, Fig. 14 (b)). A strong impact was also obtained in terms of particle. As shown in Fig. 15, particles increased with lens contamination in the case without water-addition. Some portions of lens contamination would be removed by laser irradiation and flow into the fluid as particles. Using the water-addition unit, there were no particles in the fluid made from lens contamination. It was also confirmed that the water-addition unit does not generate any particles.

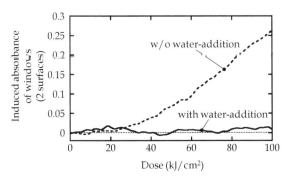

Fig. 13. Induced absorbance of windows against exposure dose. The absorbance of windows increased with the exposure dose in the case without water-addition (dashed line). In contrast, the water-addition unit suppressed contamination (solid line).

As described above, water-addition is the superior method for suppressing lens contamination. In addition, it seems to be a method without any disadvantages. For example, the refractive index of a G2 fluid does not change substantially with water-addition because the solubility of water into a G2 fluid is very low (several tens of ppm).

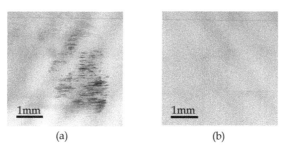

(a)   (b)

Fig. 14. Micrographs of window surfaces; (a) without water-addition after 100 kJ/cm² exposure, (b) with water-addition after 100 kJ/cm² exposure.

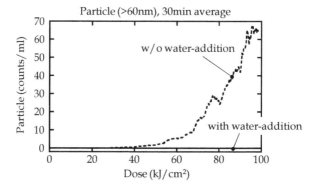

Fig. 15. Particles in a fluid against exposure dose. The number of particles increased with the exposure dose without water-addition (dashed line). In contrast, the water-addition unit suppressed particles induced by lens contamination (solid line).

3.2.4 Long-term test

A long-term test for lens contamination was done with an in-line purification unit and a water-addition unit simultaneously. IF132 (DuPont) and ARP (DuPont) were used as a G2 fluid and an in-line purification unit, respectively.

As shown in Fig. 16, the absorbance of windows started to increase at 20 kJ/cm² without any suppression units. The dose estimation on an actual exposure tool is approximately 60 kJ/(cm²·day). Therefore, lens cleaning must be done three times per day in this case. Even if with ARP, the absorbance started to rise at 60 kJ/cm² and a lens needs to be cleaned once a day. The cleaning interval is not allowed in the case with ARP alone. On the other hand, lens contamination was suppressed until 420 kJ/cm² using ARP and a water-addition unit simultaneously. The necessity for lens cleaning is once a week and it is practical. Figure 17 shows micrographs of window surfaces after the experiments. The micrographs also exhibit that the simultaneous use of ARP and a water-addition unit has a sufficient performance. It is concluded that lens contamination is not a critical issue anymore.

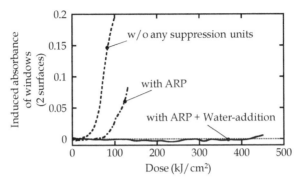

Fig. 16. Induced absorbance of windows against exposure dose. Simultaneous use of ARP and a water-addition unit suppressed lens contamination until 420 kJ/cm².

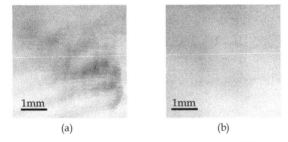

Fig. 17. Micrographs of window surfaces; (a) with ARP after 130 kJ/cm² exposure, (b) with ARP and a water-addition unit after 450 kJ/cm² exposure.

3.3 Bubble

A bubble is one of the origins of immersion defect. A bubble is entrapped on an air-liquid boundary and is translated with a wafer motion. If the bubble lifetime is sufficiently short, the bubble will be eliminated before reaching an exposure area. Therefore, the reduction of bubble lifetime is important.

The gas in a bubble diffuses from the surface of the bubble into the fluid. This diffusion causes the bubble to shrink with time and to eventually vanish. The bubble lifetime is approximated using eq. (5) (Honda et al., 2004).

$$\tau = \frac{\rho d^2}{8D(C_s - C_\infty)},\tag{5}$$

where τ is the bubble lifetime, ρ is the density of the gas inside the bubble, d is the diameter of the bubble, D is the diffusion coefficient, C_s is the saturated concentration, and C_∞ is the concentration of the dissolved gas at a position far from the bubble. Figure 18 shows a schematic diagram of the experimental setup for the bubble lifetime measurement. By increasing the pressure of the fluid from 1 to 2 atm, the bubble vanishes at a certain lifetime. It corresponds to the lifetime in the half-degassed fluid. Figure 19 shows the bubble lifetimes in water and two types of G2 fluids. The experimental results in water correlated with the theory fairly well. The bubble lifetime in the G2 fluid was shorter than that in water. Hence, the concept of the bubble elimination method for a water immersion system, i.e., using degassed fluid, is also feasible in an immersion system using a G2 fluid.

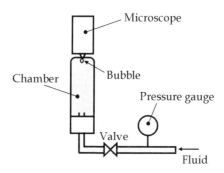

Fig. 18. Schematic diagram of the experiment setup for the bubble lifetime measurement. The size of a bubble is observed using the microscope.

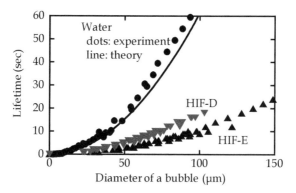

Fig. 19. Bubble lifetimes in various fluids. The lifetimes in G2 fluids (triangles) are shorter than that in water (circles).

3.4 Residual fluid on a wafer

The surface tension and viscosity of water are 72 mN/m and 1 mPa·s, respectively. This extremely high surface tension and low viscosity enable the immersion nozzle to keep water under a final lens (Kubo et al, 2007). On the other hand, it has been reported that a G2 fluid is easy to remain on a wafer (Sewell et al, 2007). After the verification that residual fluid is hard to avoid, some issues arising with residual fluid are discussed. The experimental results for the issues, such as the interaction between a G2 fluid and a resist, metal contamination in a G2 fluid, and fluid darkening due to oxygen-rich residual fluid, are explained.

3.4.1 Scanning test

The surface tension and viscosity of G2 fluids are approximately 30 mN/m and 2~4 mPa·s, respectively. The film pulling velocity (v_{fp}), which is the minimum scanning velocity to leave fluid behind the nozzle, is proportional to the surface tension (γ) and the inverse of the viscosity (μ) as shown in eq. (6) (Shedd et al., 2006). If a wafer has the same static receding contact angle (SRCA, $\theta_{s,r}$) for water and a G2 fluid, the film pulling velocity for a G2 fluid is approximately one-sixth-fold lower than that for water. Although the higher SRCA can increase the film pulling velocity, the target SRCA over 120° is too high to develop an appropriate topcoat material (Sanders et al., 2008). Therefore, it is difficult to achieve a sufficient scanning speed without residual fluid in an immersion system using a G2 fluid.

$$v_{fp} = C \frac{\gamma}{\mu} \theta_{s,r}^3 , \qquad (6)$$

where C is the empirical constant.

Figure 20 shows the results of the fundamental scanning test using a prototype topcoat for G2 fluids. HIL-203 (JSR) was used as a G2 fluid for the experiment. Even with the scanning speed of 30 mm/sec, the dynamic receding contact angle became 0° and fluid was left behind the nozzle. As a result, it was confirmed that residual fluid should be allowed to realize a sufficient scanning speed. Since the contact time of a G2 fluid and a resist becomes longer in such a system, the interaction between a G2 fluid and a resist should be discussed.

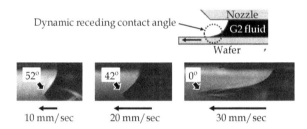

Fig. 20. Side views of the fundamental scanning test. The dynamic receding contact angle on a wafer became 0° with the scanning speed of 30 mm/sec.

3.4.2 Soaking test

To investigate the interaction between a G2 fluid and a resist, the defect inspection of soaked wafers is effective. Figure 21 shows the procedure of a soaking test. Between the exposure and post-exposure bake (PEB), a wafer was soaked in a G2 fluid for 60 seconds and the fluid was removed by spin drying. The defect inspection was carried out using KLA2371 (KLA-Tencor). The defects induced by post-exposure soaking were evaluated with the above procedure.

Figure 22 shows the results of the defect inspection. A large number of defects were found on the resist after soaking in HIF-E. In a microscopic review, there were many stains as shown in Fig. 22. As a result, it was confirmed that it is necessary to use an inert resist for a G2 fluid. On the other hand, it is not apparent that the wafer soaked in HIF-A or HIF-D had more defects than the reference wafer. There were some variations in the number of defects because the experiment was carried out in off-line process. Defect studies should be done in more clean circumstances.

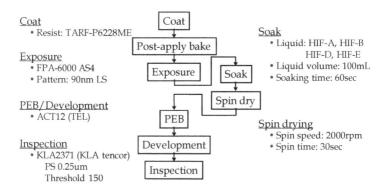

Fig. 21. Experimental procedure of a post-exposure soaking test to investigate the interaction between a G2 fluid and a resist.

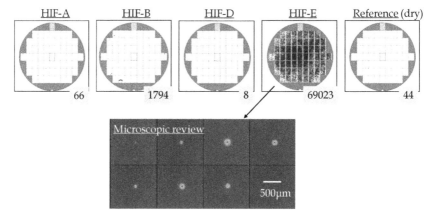

Fig. 22. Results of defect inspection. The number of defects after soaking in HIF-A, B, D, and E are 66, 1794, 8, and 69023, respectively.

3.4.3 Metal contamination

If the fluid contains a lot of metals, a wafer would be polluted by metals in the residual fluid. Thus, the objective of the following experiment is to confirm that G2 fluids are sufficiently clean. The requirement is less than 10^{10} atoms/cm^2 on a silicon wafer.

The conditions for the experiment are shown in Table 4. At first, a silicon wafer was dipped into a fluid for 5 minutes. Then, the wafer was pulled up and was dried by air blowing. Then, the metal concentrations on the wafer were evaluated by vapor phase decomposition inductively coupled plasma mass spectrometry (VPD/ICP-MS). The detection limit of the spectrometry is 0.03×10^{10} atoms/cm^2. The experiment was done twice for each fluid.

Table 5 shows the experimental results. The metal concentrations of the reference mean those on a wafer without dipping into any fluids. In the case of HIF-A, the concentrations of some metals are more than 10^{10} atoms/cm^2 (black-out cells in Table 5). On the other hand, HIF-B and D are sufficiently clean. As a result, the residual fluids of HIF-B and D are acceptable from the viewpoint of metal contamination.

	Immersion fluid	HIF-A, HIF-B, HIF-D
Dipping	Volume of fluid	200 mL
	Dipping time	5 minutes
	Wafer	Bare Si (200 mmϕ)
Analysis	Method	VPD/ICP-MS
	Evaluation area	180 mmϕ
	Detection limit	0.03×10^{10} atoms/cm^2

Table 4. Experimental conditions for metal contamination on a wafer dipped into a G2 fluid.

Fluid		Metal concentration ($\times10^{10}$atoms/cm^2)												
		Li	Na	Mg	Al	K	Ca	Ti	Cr	Mn	Fe	Ni	Cu	Zn	Pb
Refe-	1	<0.03	0.03	<0.03	0.03	<0.03	<0.03	0.43	<0.03	<0.03	<0.03	<0.03	<0.03	<0.03	<0.03
rence	2	0.06	<0.03	<0.03	0.05	<0.03	<0.03	0.32	<0.03	<0.03	<0.03	<0.03	0.03	<0.03	<0.03
HIF-A	1	0.15	0.30	0.09	1.30	0.28	0.28	0.71	0.29	0.09	1.57	0.87	0.05	0.13	<0.03
	2	0.11	0.92	0.05	0.85	0.97	0.36	0.45	0.13	0.19	0.82	0.40	0.04	0.10	<0.03
HIF-B	1	<0.03	0.45	0.05	0.23	0.13	0.38	0.48	<0.03	<0.03	0.10	<0.03	0.04	0.20	<0.03
	2	<0.03	0.48	0.07	0.29	0.14	0.47	0.58	<0.03	<0.03	0.06	<0.03	0.03	0.40	<0.03
HIF-D	1	<0.03	0.36	0.12	0.57	0.10	0.21	0.51	<0.03	<0.03	0.09	<0.03	0.07	0.06	<0.03
	2	<0.03	0.26	0.09	0.47	0.13	0.17	0.66	<0.03	<0.03	0.10	<0.03	0.04	0.03	<0.03

Table 5. Metal concentrations on a wafer evaluated by VPD/ICP-MS.

3.4.4 Dissolved oxygen

Oxygen in the atmosphere diffuses into the residual fluid on a wafer. When the oxygen-rich residual fluid goes back under the lens, it decreases the transparency of the fluid under the lens. That is the reason why dissolved oxygen is one of the issues arising with a residual fluid. Oxygen concentration in the fluid under the lens was evaluated while a wafer was moving along the sequence as shown in Fig. 23. HIL-203 (JSR) was used as a G2 fluid. A prototype immersion nozzle designed for a G2 fluid was used for the experiment. The processing time for one wafer was 30 seconds and it consisted of a scanning time and a waiting time at the edge of the stage. The fluid under the lens was sucked through the hole opened in the lens and the oxygen concentration in the sucked fluid was measured. The absorbance was calculated by using the relation as shown in Fig. 4.

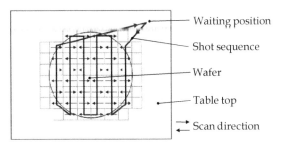

— Waiting position
— Shot sequence
— Wafer
— Table top
— Scan direction

Fig. 23. Sequence of a scanning test on an 8 inches wafer.

Figure 24 shows fluid absorbance induced by dissolved oxygen against the number of scanned waters. The absorbance rises with the scanning speed but it is low enough even at the scanning speed of 800 mm/sec. This result exhibits that the nozzle can reduce the amount of residual fluid sufficiently and can suppress the fluid darkening. In conclusion, the issue of dissolved oxygen was solved by the immersion nozzle designed for a G2 fluid.

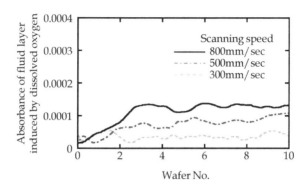

Fig. 24. Fluid absorbance between a lens and a wafer during the scan. The absorbance induced by dissolved oxygen is low enough even at the scanning speed of 800 mm/sec.

4. Remaining challenges

There are some remaining challenges to realize high-index immersion lithography. As described in the second subchapter, the quality of LuAG material does not reach the target specification. Especially, the absorbance of LuAG would be a critical issue. While the intrinsic absorbance of LuAG is lower than the target absorbance, further reduction of impurities is required.

On the other hand, there still remain a few issues to be studied in an immersion system using a G2 fluid. Although fluid absorbance can be kept low enough, dose homogeneity through a fluid layer should be confirmed. Defect study should be done with various resists. It is preferable to examine them with a preproduction tool using a G2 fluid.

For the extendability of high-index immersion lithography, the research activity on new materials such as G3 fluids and high-index resists is needed.

5. Conclusion

It has been discussed the feasibility on a high-index immersion system of 1.55 NA using LuAG and a G2 fluid. Although the IBR correction of LuAG is feasible, the quality of LuAG is not enough and the acceleration of its development is desired. The immersion system using a G2 fluid is being developed without serious issues. It was demonstrated that fluid absorbance can be kept low enough through an in-line purification unit and an oxygen removal unit. Lens contamination can be suppressed by addition of a small amount of water into a G2 fluid. Some issues arising with residual fluid, such as fluid darkening due to reentry of oxygen-rich residual fluid, were solved. By accepting residual fluid on a wafer, the scanning speed and the throughput can be raised.

EUVL is the orthodox candidate of the next generation lithography but still needs real verifications in various items. If EUVL is delayed, high-index immersion lithography will be at the leading edge lithography. It is a common knowledge that strong supports from the industry are indispensable to overcoming the remaining challenges such as LuAG. It is just said, "No market, no tool."

6. Acknowledgment

The author would like to acknowledge DuPont, JSR, and Mitsui Chemicals for supplying G2 fluids and purification units. The author also thanks colleagues for their cooperation.

7. References

Parthier, L.; Wehrhan, G.; Seifert, F.; Ansorg, M.; Aichele, T. & Seitz, C. (2008). High-index lens material LuAG : development status and progress, *presented at SEMATECH Litho Forum*, Bolton Landing, New York, USA, May 12-14, 2008

Parthier, L.; Wehrhan, G.; Seifert, F.; Ansorg, M.; Aichele, T.; Seitz, C. & Letz, M. (2008). Development update of high index lens material LuAG for ArF hyper NA immersion systems, *presented at 5th Int. Symp. on Immersion Lithography*, HI-01, The Hague, Netherlands, September 22-25, 2008

Burnett, J. H.; Kaplan, S. G.; Shirley, E. L.; Horowitz, D.; Josell, D.; Clauss, W.; Grenville, A. & Peski, C. V. (2006). High index materials for 193 nm immersion lithography, *presented at 3rd Int. Symp. on Immersion Lithography*, OO-21, Kyoto, Japan, October 2-5, 2006

Nawata, T.; Inui, Y.; Masada, I.; Nishijima, E.; Mabuchi, T.; Mochizuki, N.; Satoh, H. & Fukuda, T. (2007). High-index fluoride materials for 193 nm immersion lithography, *Proc. SPIE* Vol. 6520, 65201P

Letz, M.; Gottwald, A.; Richter, M.; Liberman, V. & Parthier, L. (2010). Temperature-dependent Urbach tail measurements of lutetium aluminum garnet single crystals, *Phys. Rev. B* 81, 155109

French, R. H.; Liberman, V.; Tran, H. V.; Feldman, J.; Adelman, D. J.; Wheland, R. C.; Qiu, W.; McLain, S. J.; Nagao, O.; Kaku, M.; Mocella, M.; Yang, M. K.; Lemon, M. F.; Brubaker, L.; Shoe, A. L.; Fones, B.; Fischel, B. E.; Krohn, K.; Hardy, D. & Chen, C. Y. (2007). High-index immersion lithography with second-generation immersion fluids to enable numerical aperatures of 1.55 for cost effective 32-nm half pitches, *Proc. SPIE* Vol. 6520, 65201O

Furukawa, T.; Kishida, T.; Miyamatsu, T.; Kawaguchi, K.; Yamada, K.; Tominaga, T.; Slezak, M. & Hieda, K. (2007). High-refractive index materials design for ArF immersion lithography, *Proc. SPIE* Vol. 6519, 65190B

Kagayama, A.; Wachi, H.; Namai, Y. & Fukuda, S. (2007). High-refractive index fluids for second-generation 193-nm immersion lithography, *persented at SPIE Advanced Lithography*, 6519-66, San Jose, California, USA, February 25-March 2, 2007

Zimmerman, P. A.; Byers, J.; Rice, B.; Ober, C. K.; Giannelis, E. P.; Rodriquez, R.; Wang, D.; O'Connor, N.; Lei, X.; Turro, N. J.; Liberman, V.; Palmacci, S.; Rothschild, M.; Lafferty, N. & Smith, B. W. (2008). Development and evaluation of a 193nm immersion generation-three fluid candidates, *Proc. SPIE* Vol. 6923, 69230A

Sakai, K.; Iwasaki, Y.; Mori, S.; Yamada, A.; Ogusu, M.; Yamashita, K.; Nishikawara, T.; Hara, S. & Watanabe, Y. (2008). Feasibility study on immersion system using high-index materials, *Jpn. J. Appl. Phys.* Vol. 47, No. 6, pp. 4853-4861

French, R. H.; Qiu, W.; Yang, M. K.; Wheland, R. C.; Lemon, M. F.; Shoe, A. L.; Adelman, D. J.; Crawford, M. K.; Tran, H. V.; Feldman, J.; McLain, S. J. & Peng, S. (2006). Second generation fluids for 193nm immersion lithography, *Proc. SPIE* Vol. 6154, 615415

Santillan, J.; Otoguro, A.; Itani, T.; Fujii, K.; Kagayama, A.; Nakayama, N.; Tamatani, H. & Fukuda, S. (2006). Novel high refractive index fluids for 193-nm immersion lithography, *Proc. SPIE* Vol. 6154, 61544Q

Sekine, Y.; Kawashima, M.; Sakamoto, E.; Sakai, K.; Yamada, A. & Honda, T. (2007). Feasibility of 37-nm half-pitch with ArF high-index immersion lithography, *Proc. SPIE* Vol. 6520, 65201Q

French, R. H.; Yang, M. K.; Peng, S.; Qiu, W.; Wheland, R. C.; Lemon, M. F.; Crawford, M. K.; Sewell, H.; McCafferty, D. & Markoya, L. (2005). Imaging of 32-nm 1:1 lines and spaces using 193-nm immersion interference lithography with second-generation Immersion fluids to achieve a numerical aperture of 1.5 and a k1 of 0.25, *J. Microlith., Microfab., Microsyst.* Vol. 4, 3, 031103

Wang, Y.; Miyamatsu, T.; Furukawa, T.; Yamada, K.; Tominaga, T.; Makita, Y.; Nakagawa, H.; Nakamura, A.; Shima, M.; Kusumoto, S.; Shimokawa, T. & Hieda, K. (2006). High-refractive-index fluids for the next generation ArF immersion lithography, *Proc. SPIE* Vol. 6153, 61530A

Furukawa, T.; Hieda, K.; Wang, Y.; Miyamatsu, T.; Yamada, K.; Tominaga, T.; Makita, Y.; Nakagawa, H.; Nakamura, A.; Shima, M. & Shimokawa, T. (2006). High refractive index fluid for next generation ArF immersion lithography, *J. Photopolym. Sci. Technol.* Vol. 19, 5, pp. 641-646

Gejo, J. L.; Kunjappu, J. T.; Zhou, J.; Smith, B. W.; Zimmerman, P.; Conley, W. & Turro, N. J. (2007). Outlook for potential third-generation immersion fluids, *Proc. SPIE* Vol. 6519, 651921

Liberman, V.; Rothschild, M.; Palmacci, S. T.; Zimmerman, P. A. & Grenville, A. (2007). Laser durability studies of high-index immersion fluids : fluid degradation and optics contamination effects, *Proc. SPIE* Vol.6520, 652035

Sakai, K.; Mori, S.; Sakamoto, E.; Iwasaki, Y.; Yamashita, K.; Hara, S.; Watanabe, Y. & Suzuki, A. (2008). Progress of high-index immersion exposure system, *presented at 5th Int. Symp. on Immersion Lithography*, HI-06, The Hague, Netherlands, September 22-25, 2008

Honda, T.; Kishikawa, Y.; Tokita, T.; Ohsawa, H.; Kawashima, M.; Ohkubo, A.; Yoshii, M.; Uda, K. & Suzuki, A. (2004). ArF immersion lithography: critical optical issues, *Proc. SPIE* Vol. 5377, pp. 319-328

Kubo, H.; Hata, H.; Sakai, F.; Deguchi, N.; Iwanaga, T. & Ebihara, T. (2007). Immersion exposure tool for the 45-nm HP mass production, *Proc. SPIE* Vol. 6520, 65201X

Sewell, H.; Mulkens, J.; Graeupner, P.; McCafferty, D.; Markoya, L.; Donders, S.; Samarakone, N. & Duesing, R. (2007). Extending immersion lithography with high index materials, *Proc. SPIE* Vol. 6520, 65201M

Shedd, T. A.; Schuetter, S. D.; Nellis, G. F. & Peski, C. V. (2006). Experimental characterization of the receding meniscus under conditions associated with immersion lithography, *Proc. SPIE* Vol. 6154, 61540R

Sanders, D. P.; Sundberg, L. K.; Brock, P. J.; Ito, H.; Truong, H. D.; Allen, R. D.; McIntyre, G. R. & Goldfarb, D. L. (2008). Self-segregating materials for immersion lithography, *Proc. SPIE* Vol. 6923, 692309

2

Laser-Plasma Extreme Ultraviolet Source Incorporating a Cryogenic Xe Target

Sho Amano
University of Hyogo
Laboratory of Advanced Science and Technology for Industry (LASTI)
Japan

1. Introduction

Optical lithography is a core technique used in the industrial mass production of semiconductor memory chips. To increase the memory size per chip, shorter wavelength light is required for the light source. ArF excimer laser light (193 nm) is used at present and extreme ultraviolet (EUV) light (13.5 nm) is proposed in next-generation optical lithography. There is currently worldwide research and development for lithography using EUV light (Bakshi, 2005). EUV lithography (EUVL) was first demonstrated by Kinoshita et al. in 1984 at NTT, Japan (Kinoshita et al., 1989). He joined our laboratory in 1995 and has since been actively developing EUVL technology using our synchrotron facility NewSUBARU. Today, EUVL is one of the major themes studied at our laboratory.

To use EUVL in industry, however, a small and strong light source instead of a synchrotron is required. Our group began developing laser-produced plasma (LPP) sources for EUVL in the mid-1990s (Amano et al., 1997). LPP radiation from high-density, high-temperature plasma, which is achieved by illuminating a target with high-peak-power laser irradiation, constitutes an attractive, high-brightness point source for producing radiation from EUV light to x-rays.

Light at a wavelength of 13.5 nm with 2% bandwidth is required for the EUV light source, which is limited by the reflectivity of Mo/Si mirrors in a projection lithography system. Xe and Sn are known well as plasma targets with strong emission around 13.5 nm. Xe was mainly studied initially because of the *debris problem*, in which debris emitted from plasma with EUV light damages mirrors near the plasma, quickly degrading their reflectivity. This problem was of particular concern in the case of a metal target such as Sn because the metal would deposit and remain on the mirrors. On the other hand, Xe is an inert gas and does not deposit on mirrors, and thus has been studied as a deposition-free target. Because of this advantage, researchers initially studied Xe. To provide a continuous supply of Xe at the laser focal point, several possible approaches have been investigated: employing a Xe gas puff target (Fiedrowicz et al., 1999), Xe cluster jet (Kubiak et al., 1996), Xe liquid jet (Anderson et al., 2004; Hansson et al., 2004), Xe capillary jet (Inoue et al., 2007), stream of liquid Xe droplets (Soumagne et al., 2005), and solid Xe pellets (Kubiak et al., 1995). Here, there are solid and liquid states, and their cryogenic Xe targets were expected to provide higher laser-to-EUV power conversion efficiency (CE) owing to their higher density compared with the gas state. In addition, a smaller gas load to be evacuated by the exhaust pump system was expected.

We have also studied a cryogenic Xe solid target. In that study, we measured the EUV emission spectrum in detail, and we found and first reported that the emission peak of Xe was at 10.8 nm, not 13 nm (Shimoura et al., 1998). This meant we could only use the tail of the Xe plasma emission spectrum, not its peak, as the radiation at 13.5 nm wavelength with 2% bandwidth. From this, improvements in the CE at 13.5 nm with 2% bandwidth became a most critical issue for the Xe plasma source; such improvements were necessary to reduce the pumped laser power and cost of the whole EUV light source. On the other hand, the emission peak of a Sn target is at 13.5 nm; therefore, Sn intrinsically has a high CE at 13.5 nm with 2% bandwidth. The CE for Sn is thus higher than that for Xe at present, in spite of our efforts to improve the CE for Xe. This resulted in a trend of using Sn rather than Xe in spite of the debris problem. Today, Cymer (Brandt et al., 2010) and Gigaphoton (Mizoguti et al., 2010), the world's leading manufacturers of LPP-EUV sources, are developing sources using Sn targets pumped with CO_2 lasers while making efforts to mitigate the effects of debris.

In the historical background mentioned above, we developed an LPP-EUV source composed of 1) a fast-rotating cryogenic drum system that can continuously supply a solid Xe target and 2) a high-repetition-rate pulse Nd:YAG slab laser. We have developed the source in terms of its engineering and investigated potential improvements in the CE at 13.5 nm with 2% bandwidth. The CE depends on spatial and temporal Xe plasma conditions (e.g., density, temperature, and size). To achieve a high CE, we controlled the condition parameters and attempted to optimize them by changing the pumping laser conditions. We initially focused on parameters at the wavelength of 13.5 nm with 2% bandwidth required for an EUV lithography source, but the original emission from the Xe plasma has a broad spectrum at 5–17 nm. We noted that this broad source would be highly efficient and very useful for many other applications, if not limiting for EUVL. Therefore, we estimated our source in the wavelength of 5–17 nm. Though Xe is a deposition-free target, there may be sputtering due to the plasma debris. We therefore investigated the plasma debris emitted from our LPP source, which consists of fast ions, fast neutrals, and ice fragments. To mitigate the sputtering, we are investigating the use of Ar buffer gas. In this chapter, we report on the status of our LPP-EUV source and discuss its possibilities.

2. Target system – Rotating cryogenic drum

We considered using a cryogenic solid state Xe target and developed a rotating drum system to supply it continuously, as shown in Fig. 1 (Fukugaki et al., 2006). A cylindrical drum is filled with liquid nitrogen, and the copper surface is thereby cooled to the temperature of liquid nitrogen. Xe gas blown onto the surface condenses to form a solid Xe layer. The drum coated with a solid Xe layer rotates around the vertical z-axis and moves up and down along the z-axis during rotation, moving spirally so that a fresh target surface is supplied continuously for every laser shot. A container wall surrounds the drum surface, except for an area around the laser focus point. This maintains a relatively high-density Xe gas in the gap between the container wall and the drum surface so as to achieve a high growth rate of the layer and fast recovery of the laser craters during rotation. The container wall also suppresses Xe gas leakage to the vacuum chamber to less than 5%, and the vacuum pressure inside the chamber is kept at less than 0.5 Pa. The diameter of the drum is 10 cm. Its mechanical rotation and up–down speed are tunable at 0–1200 rpm and 0–10 mm/s in a range of 3 cm respectively.

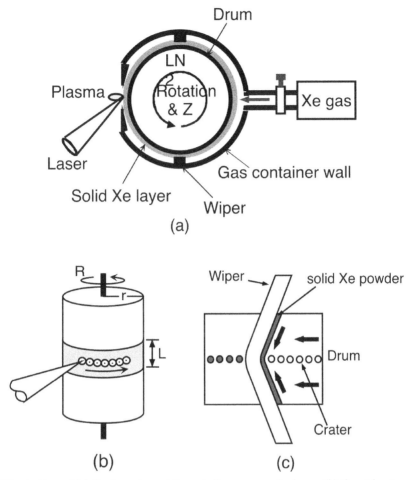

Fig. 1. Illustration of (a) the top view of the rotating cryogenic drum, (b) the side view, and (c) the wiper.

First, we formed a solid Xe layer with thickness of 300–500 µm on the drum surface and measured the size of the laser crater, which depends on the laser pulse energy. The crater diameter was measured directly from a microscope image, and its depth was roughly estimated from the number of shots needed to burn through the known thickness of the layer. A Q-switched 1064 nm Nd:YAG laser was focused on the Xe target surface with a spot diameter of 90 µm. Measured crater diameters D_c and crater depths δ_c are plotted in Fig. 2 for a laser energy range of 0.04–0.7 J. From the results in Fig. 2, a thickness of more than 200 µm was found to be sufficient for a laser shot of 1 J not to damage the drum surface. We then decided the target thickness to be 500 µm.

Two wipers are mounted on the container wall as shown in Fig.1 (a) to adjust the thickness of the solid Xe layer to 500 µm. As shown in Fig. 1 (c), the V-figure wipers also collect the Xe target powder on the craters produced by laser irradiation, thereby increasing the recovery speed. The wipers demonstrated a recovery speed of 150 μm/s up to a rotation speed of 1000 rpm, at a Xe flow rate of 400 mL/min.

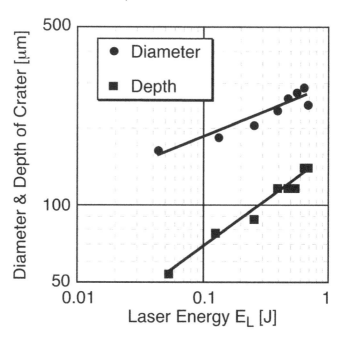

Fig. 2. Measured diameter and depth of a crater as a function of the irradiating laser energy.

Next, operational parameters of the drum are discussed to achieve high-repetition-rate laser pulse irradiation. In Fig. 1(b), R is the rotation speed, r is the radius of the drum, and L is the range of motion (scanning width of the target) along the rotational axis (z-axis). When the laser pulses are irradiated with frequency f, craters form on the target with separation length d between adjacent craters. The recovery time of a crater is T. Under the condition that craters do not overlap, f and T can be written as

$$f = \frac{2\pi r \cdot R}{d} \tag{1}$$

$$T = \frac{2\pi r L}{f \cdot d^2} \tag{2}$$

For example, if we assume laser energy of $E_L = 1$ J, a formed crater has a diameter of $D_c = 300$ μm and a depth of $\delta_c = 160$ μm, and d must be at least 300 μm for the craters not to overlap. At $r = 5$ cm and $R = 1000$ rpm, we obtain $f = 17$ kHz from Eq. (1). When $f = 10$ kHz and $L = 3$ cm, T is calculated to be 10 s using Eq. (2), and we know that a recovery speed of the crater ($V_c = \delta_c/T$) of 16 μm/s is required. Here, we have already obtained $V_c = 150$ μm/s via the wiper effect and the required speed has been achieved.

Although flaking of the target layer due to superimposition of shock and/or thermal waves produced by continuous laser pulses was a concern for high-repetition pulse operation, model experiments and calculations show that there is no problem up to 1 J per pulse and 10 kHz (Inoue et al., 2006).

From the above results, we conclude that the rotating drum system we developed can supply the target continuously, achieving the required laser irradiation of 10 kHz and 1 J, and thus realizing a high-average-power EUV light source.

3. Drive laser – Nd:YAG slab laser

High peak power and high focusability (i.e., high beam quality) are required for a driving laser to produce plasma. In addition, high average power is required for high throughput in industrial use such as EUVL. We express such a laser as a *high average and high peak brightness laser*, for which the average brightness and peak brightness are defined as average power/$(\lambda \cdot M^2)^2$ and peak power/$(\lambda \cdot M^2)^2$, respectively; we began studying such lasers in the 1990s (Amano et al, 1997,1999).

We attempted to realize a *high average and high peak brightness laser* using a solid-state Nd:YAG laser (Amano et al., 2001). The thermal-lens effect and thermally induced birefringence in an active medium are serious for such a laser; thus, thermal management of the amplifier head is more critical, and the design of the amplifier system must more efficiently extract energy and more accurately correct the remaining thermally induced wavefront aberrations in the pumping head. To meet these requirements, we developed a phase-conjugated master-oscillator-power-amplifier (PC-MOPA) Nd:YAG laser system consisting of a diode-pumped master oscillator and flash-lamp-pumped angular-multiplexing slab power-amplifier geometry incorporating a stimulated-Brillouin-scattering phase-conjugate mirror (SBS-PCM) and image relays (IR). The system design and a photograph are shown in Fig. 3. This laser demonstrated simultaneous maximum average power of 235 W and maximum peak power of 30 MW with $M^2 = 1.5$. The maximum pulse energy was 0.73 J with pulse duration of 24 ns at a pulse repetition rate of 320 pps. We therefore obtained, simultaneously, both high average brightness of 7×10^9 W/cm^2 sr and high peak brightness of 1×10^{15} W/cm^2 sr.

This peak brightness is enough to produce plasma but the average brightness needs to be higher for EUVL applications. The maximum average power is mainly limited by the thermal load caused by flash-lamp-pumping in amplifiers. The system design rules that we confirmed predicted that average output power at the kilowatt level can be achieved by replacing lamp pumping in the amplifier with laser-diode pumping. Since our work, it seems that there has been no major progress in laser engineering for such *high average and high peak brightness lasers*. Average power of more than 10 kW has been achieved in continuous-wave solid-state lasers using configurations of fibers (ex. IPG Photonics Corp.) or thin discs (ex. TRUMPF GmbH). On the other hand, for the short-pulse lasers mentioned above, the maximum average power remains around 1 kW (Soumagne et al., 2005), which is more than an order of magnitude less than the ~30 kW required for an industrial EUVL source. This is one of the reasons why CO_2 lasers have been preferred over Nd:YAG lasers as the driving laser. To further the industrial use of solid-state lasers, there needs to be a breakthrough to increase the average power.

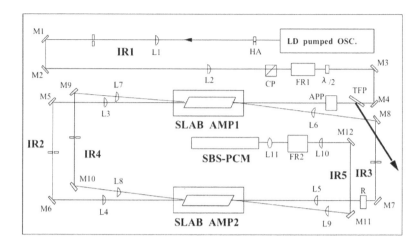

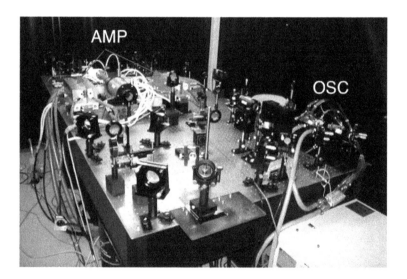

Fig. 3. Experimental setup and photograph of the PC-MOPA laser system.

4. EUV source

Figure 4 is an illustration and a photograph of the LPP-EUV source composed of a rotating cryogenic drum and Nd:YAG slab laser. The drum, detectors, and irradiating samples are installed in a vacuum chamber because EUV light cannot transmit through air. Driving laser pulses passing through the window are focused perpendicularly on the target by the lens so that Xe plasma is produced and EUV radiation is emitted. At a repetition rate of 320 Hz and average power of 110 W, the laser pulses irradiate the Xe solid target on the rotating drum with laser intensity of $\sim 10^{10}$ W/cm^2. The rotation speed is 130 rpm and the vertical speed 3 mm/s. The Xe target gas is continuously supplied at a flow rate of 400 mL/min. Under these operation conditions, we obtain continuous EUV generation with average power of 1 W at 13.5 nm and 2% bandwidth.

The driving pulse energy was determined to be 0.3 J under the optimal condition that higher CE and lower debris are simultaneously achieved, as detailed below. At present, the maximum achieved CE is 0.9% at 13.5 nm with 2% bandwidth for the optimal condition. Under drum-rotating operation, we found the good characteristics of increased CE and less fast ions compared with the case with the drum at rest. We next detail the EUV and debris characteristics of the EUV source.

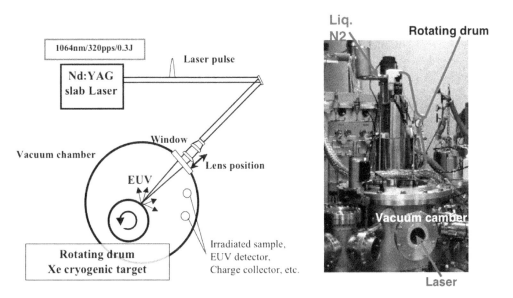

Fig. 4. Experimental setup and photograph of the laser plasma EUV source.

5. Conversion efficiency for EUVL

In this section, we report our studies carried out to improve the CE at 13.5 nm with 2% bandwidth required for the EUVL source (Amano et al., 2008, 2010a). To achieve the highest CE, we attempted to control the plasma parameter by changing the driving laser conditions. We investigated dependences of the CE on the drum rotation speed, laser energy, and laser wavelength. We also carried out double-pulse irradiation experiments to improve the CE.

To obtain data of EUV emission, a conventional Q-switched Nd:YAG rod laser (Spectra-Physics, PRO-230) was used in single-shot operation. By changing the position of the focusing lens to change the laser spot, the laser intensity on the target was adjusted to find the optimum intensity. We note that the lens position (LP) is zero at best focus, negative for in-focus (the laser spot in the target before the focus) and positive for out-of-focus (beyond the focus).

Figure 5(a) shows the CE per solid angle as a function of LP (laser intensity), which was measured by an EUV energy detector calibrated absolutely — *Flying Circus* (SCIENTEC Engineering) — located 45 degrees from the laser incident axis. The laser pulse energy was 0.8 J. We see that the CE was higher under the rotating-drum condition than under the rest condition. Here, the rest condition is as follows. Xe gas flow is stopped (0 mL/min) after the target layer has formed, and the drum rests (0 rpm) during a laser shot and stepwise rotates after every shot so that a fresh target is supplied to the point irradiated by the laser. The rotation condition is as follows. Laser pulses irradiate quasi-continuously the target on the rotating drum (>3 rpm), supplying Xe gas (>40 mL/min) and forming the target layer. The EUV intensity increased immediately with slow rotation (>3 rpm) and appeared to be almost independent of the rotation speed. In Fig. 5(a), we see that the maximum CE per solid angle was for an optimized laser intensity of 1×10^{10} W/cm^2 (LP = –10 mm) during rotation. The EUV angular distribution could be expressed by a fitting curve of $(\cos\theta)^{0.38}$, and taking into account this distribution, we obtained the maximum spatially integrated CE of 0.9% at 13.5 nm with 2% bandwidth. EUV spectra at laser intensity of 1×10^{10} W/cm^2 are shown in Fig. 5(b). It is obvious that the emission of the 13.5 nm band was greater in the case of rotation than it was in the case of rest.

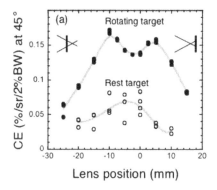 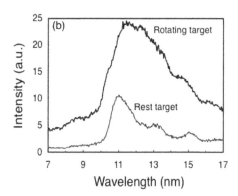

Fig. 5. (a) CE at the wavelength of 13.5 nm with 2% bandwidth as a function of LP under the rotation (130 rpm) and at-rest (0 rpm) conditions. The laser energy was 0.8 J. Insets show the laser beam focusing on the target. (b) Spectra of EUV radiation from the cryogenic Xe drum targets under the rotation (bold line) and at-rest (narrow line) conditions with laser intensity of 1×10^{10} W/cm^2 for LP of –10 mm.

We considered the mechanism for the increase in EUV intensity with rotation of the target. Figure 6 shows photographs of the visible emission from the Xe target observed from a transverse direction. It shows an obvious expansion of the emitting area with longer (optically thicker) plasma in the rotating case compared with the at-rest case. These images indicate the existence of any gas on the target surface. Under the rotation condition, Xe gas is supplied continuously to grow the target layer and the wipers form the layer. However, the wipers are not chilled especially, and the temperature of the target surface might increase owing to contact with the wipers in the rotating case so that the vapor pressure increases. Therefore, the vaporized Xe gas from the target surface was considered as the gas on the target. Although additional Xe gas was added from outside the vacuum chamber, the EUV intensity did not increase and in fact decreased owing to gas absorption. Therefore, it is supposed that Xe gas with adequate pressure localizes only near the target surface. From these results, we conclude that Xe gas on the target surface in the rotating drum produces optically thick plasma that has optimized density and temperature for emitting EUV radiation, and satellite lines of the plasma contribute effectively to increasing the EUV intensity (Sasaki et al., 2004).

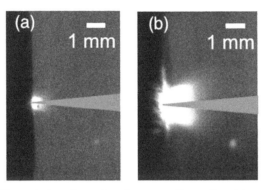

Fig. 6. Images of visible emissions from the plasma on the resting (a) and rotating (b) targets.

Next, the dependence of the laser pulse energy was investigated. We measured the CE as a function of laser energy at different LPs in the rotating drum. For laser energies exceeding 0.3 J, a CE of nearly 0.9% was achieved by tuning the LP with the laser intensity optimized as ~10^{10} W/cm². In the energy range, the maximum CE did not depend on the laser energy. At the LP in this experiment, the spot size on the target was larger than 500 μm and plasma energy loss at the edges could be ignored for this large spot. Therefore, the same CE was achieved at the same laser intensity. However, in the lower energy region, the spot size must be small to achieve optimal laser intensity, and edge loss due to three-dimensional expansion in plasma cannot then be ignored and a decrease in the CE was observed. Therefore, it is concluded that laser energy must exceed 0.3 J to achieve a high CE.

The dependence of the laser wavelength was also investigated. Additionally, we carried out 1 ω double-pulse irradiation experiments in which a pre-pulse produces plasma with optimal density and temperature, and after a time delay, a main laser pulse effectively injects emission energy into the expanded plasma to increase the CE. Under the rest condition, there were increases in CE for the shorter laser or the double pulse irradiation (Miyamoto et al., 2005, 2006). In both cases, the long-scale plasmas and their emission spectra were observed to be similar to those under the rotation condition for 1 ω single-pulse irradiation. Therefore, we supposed that in the both cases, the CE was increased by the same mechanism described above. However, when the shorter pulses or the double pulses were emitted under the rotating condition, the CE did not increase but decreased. It is considered that the opacity of the plasma was too great in these experiments and the best condition was not achieved.

In conclusion, the maximum CE was found to be 0.9% at 13.5 nm with 2% bandwidth for the optimal condition.

6. Xe plasma debris

In this section, we report the characteristics of the plasma debris that damages mirrors (Amano et al., 2010b). First, we investigated fast ions, fast neutrals and ice fragments, which constitute the debris.

When we found that EUV radiation was greater for a rotating drum than for a drum at rest, we also found that the number of fast ions decreased simultaneously. Figure 7(a) shows ion signals from a charge collector (CC) with laser pulse energy of 0.5 J and optimal intensity of 10^{10} W/cm², for different drum rotation speeds. The ion signal reduces rapidly after the drum starts to rotate (> 4 rpm), after which the signal is almost independent of rotation speed. Ion energy spectra were obtained as shown in Fig. 7(b) using the time-of-flight signals shown in Fig. 7(a). Here, we assume that all ions were doubly charged because we measured the principle charge state of Xe ions to be two with an electrostatic energy analyzer (Inoue et al., 2005). Under the rotation condition, the maximum ion energy decreases to 6 keV and the number of high-energy ions (with energy of a few dozen kilo-electron-volts) also decreases. These are favorable characteristics for the debris problem. The decrease in the ion count under the rotation condition can be explained by a *gas curtain effect* that originates from the Xe gas localized at the target surface. The pressure of this localized Xe gas can be roughly estimated from the peak attenuation (η) in Fig. 7(a); we estimated the product of pressure and thickness to be about 10 Pa·mm.

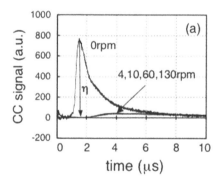
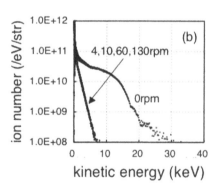

Fig. 7. (a) CC signals of ions and (b) their energy spectra at rotation speeds of 0, 4, 10, 60 and 130 rpm. η in (a) is the loss rate of ions due to the drum rotating. The ion number in (b) was calculated assuming the charge state was two.

Fast neutral particles were measured by the microchannel plate (MCP) detector when the number of fast ions decreased under the rotation condition. The MCP is sensitive to both ions and neutrals, making the use an electric field obligatory to repel ions so that the MCP detects only neutral particles. From the measurement, we found the number of neutrals to be approximately an order of magnitude less than the number of ions.

In the case of solid Xe targets, ice fragments might be produced by shock waves of laser irradiation, whereas this is not the case for gas or liquid targets. In early experiments using a solid Xe pellet, ice fragments were observed and mirror damage due to these fragments was indicated (Kubiak et al., 1995). Since these reports, liquid Xe targets have been preferred over solid Xe targets, with the exception of our group. It is therefore necessary to clarify characteristics of fragment debris from a solid Xe target on a rotating cryogenic drum. After exposing a Si sample to the Xe plasmas pumped by 100 laser pulses, we observed fragment impact damage on its surface using a scanning electron microscope. We observed damage spots on the samples at laser energy of 0.8 J irrespective of whether the drum rotates. Conversely, we did not observe spots at laser energy of 0.3 J. To explain these results, we consider that the fragment speed (kinetic energy) might drop below a damage threshold upon reducing the laser pulse energy because the fragment speed is a function of incident laser energy (Mochizuki et al., 2001). Observing the damage spots, we know that the fragment size was larger than a few microns, and the gas curtain might not be effective for such large fragments. This would explain why the fragment impact damage was independent of the state of drum rotation. From these results, we conclude that fragment impact damage, which occurs especially for the solid Xe target, can be avoided simply by reducing the incident laser pulse energy to less than 0.3 J.

The laser pulse energy was set to 0.3 J to avoid fragment impact damage and the laser repetition rate was 320 pps, giving an average power of 100 W. Next, we investigated damage to a Mo/Si mirror, which was the result of total plasma debris (mainly fast ions) from the laser multi-shots experiments. After 10 min plasma exposure, the sputtered depth was measured to be 50 nm on the surface of a Mo/Si mirror placed 100 mm from the plasma at a 22.5-degree angle to the incident laser beam. Because a typical Mo/Si mirror has 40 layer pairs and the thickness of one pair is approximately 6.6 nm, all layers will be removed within an hour by the sputtering. Although Xe is a deposition-free target, sputtering by debris needs to be mitigated. However, the major plasma debris component is ions, and we believe their mitigation to be simple compared with the case of a metal target such as Sn, using magnetic/electric fields and/or gas. We are now studying debris mitigation by Ar buffer gas. Ar gas was chosen because of its higher stopping power for Xe ions and lower absorption of EUV light, and its easy handling and low cost. After the vacuum chamber was filled with Ar gas, total erosion rates were measured using a gold-coated quartz crystal microbalance sensor placed 77 mm from the plasma at a 45-degree angle, and simultaneously, EUV losses were monitored by an EUV detector placed 200 mm from the plasma at a 22.5-degree angle. Figure 8 shows the erosion rates as a function of Ar gas pressure. The rates were normalized by the erosion N_0 at a pressure of 0 Pa. When the Ar pressure was 8 Pa, we found the erosion rate was 1/18 of that without the gas, but the absorption loss for EUV light was only 8%. The erosion rates (N/N_0) in Fig. 8 can be fitted to an exponential curve:

$$N\left(P_{Ar}\right) = N_0 \cdot \exp\left(-\frac{P_{Ar}}{kT}\sigma l\right) \qquad (3)$$

where P_{Ar} is the Ar pressure, k is the Boltzmann constant, T is the gas temperature, σ is the cross section and l is the debris flight length. From this fitting, we obtain $\sigma = 2.0 \times 10^{-20}$ m². The Ar buffer gas successfully mitigated the effect of plasma debris with little EUV attenuation. Increasing the Ar pressure, mirror erosion decreases but EUV attenuation increases. Compromising the erosion and EUV attenuation, an optimized pressure is achieved. We should localize the higher density Ar gas to only the debris path so that EUV attenuation is as small as possible. We can design the optimized pressure condition using the σ value obtained and we consider the use of an Ar gas jet. Through this mitigation, we expect that erosion will be reduced by more than two orders of magnitude and the lifetime of the mirror will be extended. We believe the debris problem for Xe plasma will thus be solved.

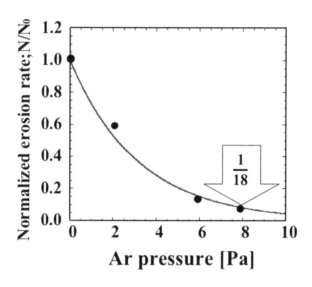

Fig. 8. Normalized erosion rate as a function of Ar pressure. The laser energy was 0.3 J and the rotation speed was 130 rpm.

7. EUV emission at 5-17nm

We began developing the LPP source for EUVL and characterized it at 13.5 nm with 2% bandwidth, but Xe plasma emission has originally a broad continuous spectrum as shown in Fig. 9. If the broad emission is used, our source will be very efficient, not limiting its applications to EUVL. We characterized the source again in the wavelength range of 5–17 nm. Figure 10 shows the CE at 5–17 nm as a function of LP (laser intensity) with laser energy of 0.8 J. The maximum spatially integrated CE at 5–17 nm was 30% for optimal laser intensity of 1×10^{10} W/cm². The maximum CE depended on the laser energy and was 21% at 0.3 J. Therefore, high average power of 20 W at 5–17 nm has been achieved for pumping by the slab laser with 100 W (0.3 J at 320 pps). We consider this a powerful and useful source.

Recently, new lithography using La/B₄C mirrors having a reflectivity peak at 6.7 nm was proposed as a next-generation candidate following EUVL using Mo/Si mirrors having a reflectivity peak at 13.5 nm (Benschop, 2009). This means that a light source emitting around 6 nm will be required in a future lithograph for industrial mass production of semiconductors. Because our source emits broadly at 5–17 nm as mentioned above, it can obviously be such a 6 nm light source. We thus next characterized it as a source emitting at 6.7 nm. Here we did not carry out new experiments to optimize the plasma for emitting at 6.7 nm but looked for indications of strong emission at 6.7 nm from the spectrum data already acquired. When making efforts to improve the CE at 13.5 nm, we noticed that emissions around 6 nm became strong at higher laser intensity. When laser energy is 0.8 J and LP = 0 mm (i.e., laser intensity is 4×10^{12} W/cm^2 under the rotation condition), there is a hump around 6 nm as shown in Fig. 9. The spatially integrated CE at 6.7 nm with 0.6% bandwidth is estimated to be 0.1% from this spectrum. Because the bandwidth of 0.6% for the La/B$_4$C mirror reflectivity is narrower than the 2% for the Mo/Si mirror, the available reflected power is intrinsically small. The CE of 0.1% was not obtained under optimized conditions and higher CE may be achieved in the future. In any event, our source is only one LPP source at present that can generate continuously an emission at 6.7 nm.

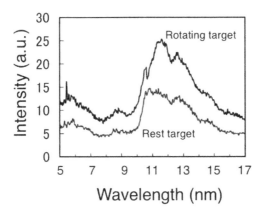

Fig. 9. Spectra of EUV radiation under the rotation (bold line) and at-rest (narrow line) conditions with laser intensity of 4×10^{12}W/cm^2 for best focus (LP = 0 mm). The laser energy was 0.8 J.

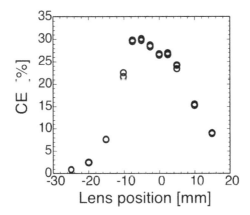

Fig. 10. CE for a wavelength of 5–17 nm as a function of LP under the rotation (130 rpm) condition. The laser energy was 0.8 J.

8. Conclusion

This chapter briefly reviewed our LPP-EUV source. First, we characterized the source at a wavelength of 13.5 nm with 2% bandwidth as an EUVL source and achieved a maximum CE of 0.9%. When the driving laser power is 110 W at 320 pps, the average power of 1 W is obtained at the wavelength and this is thought to be sufficient for the source to be used in various studies. However, the EUV power required for industrial semiconductor products is more than 100 W at present; our power is two orders of magnitude less. To approach the requirements of an industrial EUV source, the remaining tasks are considered. The majority of Xe plasma debris is fast ions, which can be mitigated using gas and/or a magnetic/electric field relatively easily. The drum system can supply the Xe target for laser pulses with energy up to 1 J at 10 kHz. Therefore, a remaining task is powering up the driving laser. A short pulse laser with average power of the order of 10 kW (i.e., *high average and high peak brightness laser*) must be developed and such a breakthrough is much hoped for.

Not limiting the wavelength to 13.5 nm with 2% bandwidth and using the broad emission at 5-17 nm, a maximum CE of 30% is achieved. Pumping with laser power of 100 W, high average power of 20 W is already obtained and the source is useful for applications other than industrial EUVL using Mo/Si mirrors. We are now applying our source to microprocessing and/or material surface modification. Our source also emits around the wavelength of 6 nm considered desirable for the next lithography source. In conclusion, our LPP source is a practicable continuous EUV source having possibilities for various applications.

9. Acknowledgment

Part of this work was performed under the auspices of MEXT (Ministry of Education, Culture, Sports, Science and Technology, Japan) under the contract subject "Leading Project for EUV lithography source development".

10. References

Amano, S., Shimoura, A., Miyamoto, S. & Mochizuki, T. (1997). High-repetition-rate pulse Nd:YAG slab laser for x-ray source by cryogenic target, *1997 OSA Technical Digest Series, Vol.11, Conference Edition, CLEO97*, p.523, Baltimore, USA, May 18-23, 1997

Amano, S., Shimoura, A., Miyamoto, S. & Mochizuki, T. (1999). Development of a high repetition rate Nd:YAG slab laser and soft X-ray generation by X-ray cryogenic target. *Fusion Eng. and Design* 44, pp.423-426

Amano, S. & Mochizuki, T. (2001). High average and high peak brightness slab laser. *IEEE J. Quantum Electron.* 37(2), pp.296-303

Amano, S., Nagano, A.: Inoue, T., Miyamoto, S. & Mochizuki, T. (2008). EUV light sources by laser-produced plasmas using cryogenic Xe and Li targets. *Rev. Laser. Eng.* 36(11), pp.715-720 (in Japanese)

Amano, S., Masuda, K., Shimoura, A., Miyamoto, S. & Mochizuki, T. (2010a). Characterization of a laser-plasma extreme ultraviolet source using a rotating cryogenic Xe target. *Appl. Phys.B* 101, pp.213-219

Amano, S., Inaoka, Y., Hiraishi, H., Miyamoto, S. & Mochizuki, T. (2010b). Laser-plasma debris from a rotating cryogenic-solid-Xe target. *Rev. Sci. Instrum.* 81, pp. 023104-1-023104-6

Anderson, R.J., Buchenauer, D.A., Klebanoff, L., Wood II, O.R. & Edwards, N.V. (2004). The erosion of materials exposed to a laser-pulsed plasma (LPP) extreme ultraviolet (EUV) illumination source, *Proceedings of SPIE, Emerging Lithographic Technologies VIII*, vol.5374, pp.710-719, Santa Clara, USA, February 2004

Bakshi,V.(Ed). (2005) *EUV Sources for Lithography*, SPIE, ISBN:0819458457, Bellingham, WA

Benschop, J. (2009). EUV: past, present and prospects, Keynote I, *2009 International Symposium on Extreme Ultraviolet Lithography*, Prague Czech, October 2009, International Sematech, Available from: <http://www.sematech.org/meetings /archives/litho/index.htm>

Brandt, D.C., Fomenkov, I.V., Partlo, W.N., Myers, D.W., Kwestens, P., Ershov, A.I., Bowering, N.R., Baumgart, P., Bykanov, A.N., Vaschenko, G.O., Khodykin, O.V., Srivastava, S., Hou, R., Dea, S.D., Ahmad, I. & Rajyuguru, C. (2010). LPP EUV source production for HVM, SO-01, *2010 International Symposium on Extreme Ultraviolet Lithography*, Kobe Japan, October 2010, International Sematech, Available from: <http://www.sematech.org/meetings/archives/litho/index.htm>

Fiedorowicz, H., Bartnik, A., Szczurek, M., Daido, H., Sakaya. N., Kmetik, V., Kato, Y., Suzuki, M., Matsumura, M., Yajima, J., Nakayma, T. & Wilhein, T. (1999). Investigation of soft X-ray emission from a gas puff target irradiated with a Nd:YAG laser, *Opt.Comm.*,163(1-3), pp.103-114

Fukugaki, K., Amano, S., Shimoura, A., Inoue, T., Miyamoto, S. & Mochizuki, T. (2006). Rotating cryogenic drum supplying Xe target to generate extreme ultraviolet radiation. *Rev. Sci. Instrum.* 77, pp.063114-1-063114-4

Hansson, B.A.M., Hemberg, O., Hertz, M.H., Berglund, M., Choi, H. J., Jacobsson, B., Janin, E., Mosesson, S., Rymell, L., Thoresen, J. & Wilner, M. (2004). Characterization of a liquid-xenon-jet laser-plasma extreme-ultraviolet source. *Rev.Sci.Instrum.* 75(6), pp.2122-2129

Inoue, T., Kaku, K., Shimoura, A., Nica, P.E., Sekioka, T., Amano, S., Miyamoto, S. & Mochizuki, T. (2005). Studies on laser-produced plasma EUV generation by using fast-supplyunig cryogenic Xe targets, 1-SO-10, *2005 International Symposium on Extreme Ultraviolet Lithography*, San Diego, USA, November 2005, International Sematech, Available from: <http://www.sematech.org/meetings/archives/litho/index.htm>

Inoue, T., Amano, S., Miyamoto, S. & Mochizuki, T. (2006). The stability of a rotating-drum solid-Xe target subjected to high-repetition rate laser irradiation for laser-plasma EUV generation. *Rev. Laser. Eng.* 34(8), pp.570-574 (in Japanese)

Inoue, T., Okino, H., Nica, P.E., Amano, S., Miyamoto, S. & Mochizuki, T. (2007). Xe capillary target for laser-plasma extreme ultraviolet source. *Rev. Sci. Instrum.* 78, pp.105105-1-105105-5

Kinoshita, H., Kurihara, K., Ishii,Y. & Torii, Y. (1989). Soft x-ray reduction lithography using multilayer mirrors. *J. Vac. Sci. Technol.B*, 7(6), pp.1648-1651

Kubiak, G., Krentz. K., Berger, K., Trucano, T., Fisher, P. & Gouge, M. (1995). Cryogenic pellet laser plasma source targets, *OSA Proceedings on Soft X-ray Projection Lithography*, vol.23, pp.248-254, Monterey, USA, September 1994

Kubiak, G., Bernardez, L.J., Krenz, K.D., O'Connell, D.J., Gutowski, R. & Todd, A.M., (1996). Debris-free EUVL sources based on gas jets, *OSA TOPS on Extreme Ultraviolet Lithography*, vol.4, pp.66-71

Miyamoto, S., Shimoura, A., Amano, S., Fukugaki, K., Kinugasa, H., Inoue, T. & Mochizuki, T. (2005). Laser wavelength and spot diameter dependence of extreme ultraviolet conversion efficiency in ω, 2ω, and 3ω Nd:YAG laser-produced plasmas. *Appl. Phys. Lett.* 86(26), pp.261502-1-261502-3

Miyamoto, S., Amano, S.: Inoue, T., Nica, P. E., Shimoura, A., Kaku, K., Sekioka, T. & Mochizuki, T. (2006). EUV source developments on laser-produced plasmas using cryogenic Xe and Lithium new scheme target, *Proceedings of SPIE, Emerging Lithographic Technologies X*, vol.6151, pp.61513S-1-61513S-10, San Jose, USA, February 2006

Mizoguti, H., Abe, T., Watanabe, Y., Ishihara, T., Ohta, T., Hori, T., Kurosu, A., Komori, H., Kakizaki, K., Sumitani, A., Wakabayashi, O., Nakarai, H., Fujimoto, J. & Endo, A. (2010). 1st generation laser-produced plasma 100W source system for HVM EUV lithography, SO-03, *2010 International Symposium on Extreme Ultraviolet Lithography*, Kobe Japan, October 2010, International Sematech, Available from: <http://www.sematech.org/meetings/archives/litho/index.htm>

Mochizuki, T., Shimoura, A., Amano, S. & Miyamoto, S. (2001). Compact high-average-power X-ray source by cryogenic target, *Proceedings of SPIE, Applications of X Rays Generated from Lasers and Other Bright Sources II*, vol.4504, pp.87-96, San Diego, USA, July 2001

Sasaki, A., Nishihara, K., Murakami, M., Koike, F., Kagawa, T., Nishikawa, T., Fujima, K., Kawamura, T. & Furukawa, H. (2004). Effect of the satellite lines and opacity on the extreme ultraviolet emission from high-density Xe plasmas. *Appl. Phys. Lett.* 85(24), pp.5857-5859

Shimoura, A., Amano, S., Miyamoto, S. & Mochizuki, T. (1998). X-ray generation in cryogenic targets irradiated by 1 μm pulse laser. *Appl. Phys. Lett.* 72(2), pp.164-166

Soumagne, G.: Abe, T., Suganuma,T. , Imai, Y., Someya, H., Hoshino, H., Nakano, M., Komori, H., Takabayashi, Y., Ariga, T., Ueno, Y., Wada, Y., Endo, A & Toyoda, K.(2005). Laser-produced-plasma light source for EUV lithography, *Proceedings of SPIE, Emerging Lithographic Technologies IX*, vol.5751, pp.822-828, San Jose, USA, March 2005

3

Irradiation Effects on EUV Nanolithography Collector Mirrors

J.P. Allain
Purdue University
United States of America

1. Introduction

Exposure of collector mirrors facing the hot, dense pinch plasma in plasma-based EUV light sources to debris (fast ions, neutrals, off-band radiation, droplets) remains one of the highest critical issues of source component lifetime and commercial feasibility of nanolithography at 13.5-nm. Typical radiators used at 13.5-nm include Xe, Li and Sn. Fast particles emerging from the pinch region of the lamp are known to induce serious damage to nearby collector mirrors. Candidate collector configurations include either multi-layer mirrors (MLM) or single-layer mirrors (SLM) used at grazing incidence. Due to the strong absorbance of 13.5-nm light only reflective optics rather than refractive optics can work in addition to the need for ultra-high vaccum conditions for its transport.

This chapter presents an overview of particle-induced damage and elucidates the underlying mechanisms that hinder collector mirror performance at 13.5-nm facing high-density pinch plasma. Results include recent work in a state-of-the-art in-situ EUV reflectometry system that measures real time relative EUV reflectivity (15-degree incidence and 13.5-nm) variation during exposure to simulated debris sources such as fast ions, thermal atoms, and UV radiation (Allain et al., 2008, 2010). Intense EUV light and off-band radiation is also known to contribute to mirror damage. For example off-band radiation can couple to the mirror and induce heating affecting the mirror's surface properties. In addition, intense EUV light can partially photoionize background gas used for mitigation in the source device. This can lead to local weakly ionized plasma creating a sheath and accelerating charged gas particles to the mirror surface inducing sputtering. In this overview we will also summarize studies of thermal and energetic particle exposure on collector mirrors as a function of temperature simulating the effects induced by intense off-band and EUV radiation found in EUVL sources. Measurements include variation of EUV reflectivity with mirror damage and in-situ surface chemistry evolution.

In this chapter the details from the EUV radiation source to the collector mirror are linked in the context of mirror damage and performance (as illustrated in Figure 1). The first section summarizes EUV radiation sources and their performance requirements for high-volume manufacturing. The section compares differences between conventional discharge plasma produced (DPP) versus laser plasma produced (LPP) EUV light sources and their possible combinations. The section covers the important subject of high-density transient plasmas and their interaction with material components. The different types of EUV radiators, debris distribution, and mitigation sources are outlined. The second section summarizes the various optical collector mirror geometries used for EUV lithography. A brief discussion on the intrinsic damage mechanisms linked to their geometry is included. The third section summarizes in general irradiation-driven mechanisms as background for the reader and its relation to the "quiescent" plasma collector mirrors are exposed in EUV sources. This includes irradiation-driven nanostructures, sputtering, ion mixing, surface diffusion, and ion-induced surface chemistry. The fourth section briefly discusses EUV radiation-driven plasmas as another source of damage to the mirror. These plasmas are a result of using gases for debris mitigation. The fifth section is a thorough coverage of the key irradiation-driven damage to optical collector mirrors and their performance limitations as illustrated in part by Figure 1.

2. EUV radiation sources

There are numerous sources designed to generate light at the extreme ultraviolet line of 13.5-nm. Historically advanced lithography has considered wavelength ranges from hard X-rays up to 157 nm [Bakshi, 2009]. Radiators of 13.5-nm light rely on high-density plasma generation typically based on discharge-produced configurations with magnetically confined high-density plasmas or laser-produced plasmas. Recently, some sources have combined both techniques (Banine 2011). Generation of high-density plasmas to yield temperatures of the order of 10-50 eV require advanced materials for plasma-facing components in these extreme environments in particular discharge-produced plasma (DPP) configurations. This is due to the need of metallic anode/cathode components operating under high-heat flux conditions. Laser-produced plasmas (LPP) benefits from the fact that no nearby electrodes are necessary to induce the plasma discharge. Further details will be described in section 5.1. One challenge in operating EUV lamps at high power is the collected efficiency of photons at the desired exposure wavelength of 13.5-nm. This particular line has a number of radiators with properties that have consequences on EUV source operation. For example radiators at 13.5-nm include xenon, tin and lithium. The latter two are metals and thus their operation complicated by contamination issues on nearby material components such as electrodes and collector mirrors. Further discussion follows in section 2.2 and 2.3. To contend with the various types of debris that are generated in the plasma-producing volume a variety of novel debris mitigation systems (DMS) have been designed and developed for both DPP and LPP configurations.

2.1 Function and material components
The transient nature of the high-density plasma environment in DPP and LPP systems results in exposure of plasma-facing components to extreme conditions (e.g. high plasma density ($\sim 10^{19}$ cm^{-3}) and temperature ($\sim$ 20-40 eV). However, in LPP systems since the configuration is mostly limited by the mass of the radiator and the laser energy supplied to it to generate highly ionized plasma with the desired 13.5-nm light. Both configurations rely on efficient radiators of 13.5-nm light, which include: Li, Sn and Xe. In DPP designs a variety of configurations have been used that include: dense plasma focus, capillary Z-pinch, star pinch, theta pinch and hollow cathode among others. For a more formal description of these high-density plasma sources for 13.5-nm light generation the author refers to the recent publications by V. Bakshi in 2006 and 2009 (Bakshi, 2006; Bakshi, 2009).

The in-band and off-band radiation generated in these sources is also a critical limitation in operation of these lamps since on average the off-band radiation is converted into heat on nearby plasma-facing components. There are additional challenges in the design of 13.5-nm light sources that include: high-frequency operation limits driven by the need to extract high EUV power at the intermediate focus (IF) and limited by the available high-throughput power of the plasma device (e.g. laser system or discharge electrode system). Additionally, the scaling of debris with EUV power extraction and the limitation of conversion efficiency (CE) with source plasma size also translate into significant engineering challenges to the design of 13.5-nm lithography source design. Figure 1 illustrates, for the case of the DPP configuration, the primary debris-generating sources that compromise 13.5-nm collector mirrors. The first region depicted on the left is defined here as the "transient plasma region". This is the region described earlier with high-density and high-temperature plasma interacting with the electrode surfaces.

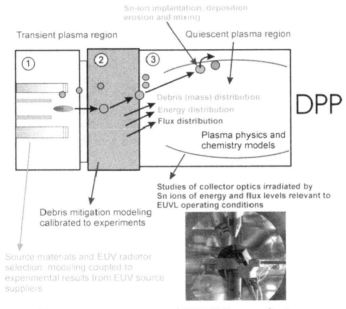

Fig. 1. Illustration of the various components of EUV 13.5-nm radiation source configuration consisting primarily of three major components: 1) plasma radiator section, 2) debris mitigation system and 3) optical collector mirror.

In DPP discharge sources material components that make up the electrode system consist of high-temperature, high-toughness materials. Although DPP source design has traditionally used high-strength materials such as tungsten and molybdenum alloys, the extreme conditions in these systems limit the operational lifetime of the electrode. Significant plasma-induced damage is found in the electrode surfaces, which induce degradation and abrasion over time. Figure 2, for example, shows a scanning electron micrograph of a tungsten electrode exposed to a dense plasma focus high-intensity plasma discharge. The key feature in the SEM image is the existence of plasma-induced damage domains that effectively have induced melting in certain sections of the electrode surface.

The second region depicted in Figure 1 is defined as the debris mitigation zone (DMZ). In this region a variety of debris mitigation strategies can be used to contend with the large debris that exists in operation of the DPP source. For example the use of inert gas to slow-down energetic particles that are generated in the pinch plasma region and/or debris mitigation shields that collect macro-scale particulates when using Sn-based radiators in DPP devices. Radiation-induced mechanisms on the surfaces of the DMZ elements also can lead to ion-induced sputtering of DM shield material that eventually is deposited in the nearby 13.5-nm collector mirror. Therefore care is taken to select sputter-resistant materials for the DM shields used such as refractory metal alloys and certain stainless steels. Design of DM shields also involve computational modeling that can aid in identifying appropriate materials depending on the source operation and generation of a variety of debris types such as clusters, ions, atoms, X-rays, electrons and macroscopic dust particles.

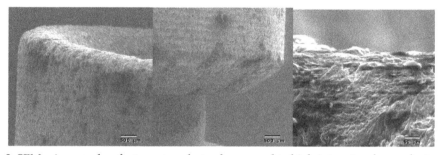

Fig. 2. SEM micrographs of a tungsten electrode exposed to high-intensity plasma during the generation of EUV 13.5-m light.

The third region in Fig. 1 consists of the 13.5-nm light collector mirror. The collector mirror has a configuration to optimally collect as much of the 13.5-nm light as possible. Its function is to deliver EUV power in a specified etendue at the intermediate focus (IF) or the opening of the illuminator. This power is in turn dictated by the specification on EUV exposure of the EUV lithography scanner that must be able to operate with 150-200 wafers per hour (wph) at nominal power for periods of 1-2 years without maintenance (so-called high-volume manufacturing, HVM, conditions). This ultra-stringent requirement is one of the primary challenges to EUV lithography today. Since powers of order 200-300 W at the IF need to be sustained for a year or more, materials at the DPP source and those used for collector mirrors will necessarily require revolutionary advances in materials performance. The third region in Figure 1 also depicts what debris the collector mirror is exposed to during the discharge. A distribution of debris energies (i.e. ions), fluxes and masses will effectively affect the mirror surface performance. The third region is also known as the "condenser or collector optics region".

2.2 Selection of electrode materials in DPP EUV devices
Selection of materials for DPP electrodes depends on the microstructure desired to minimize erosion and maximize thermal conductivity. Figure 3 shows an example of SEM micrographs of materials identified to have promising EUV source electrode properties. The powder composite materials inherited the structural characteristics of the initial powders, determined by the processes of combined restoration of tungsten and nickel oxides (WO_3

and NiO from $NiCO_3$, for instance) and copper molybdate ($MoCuO_4$). Dry hydrogen (the dew point temperature is above 20 °C) facilitates the formation of the heterogeneous conglomerates in W-Ni-powders, which do not collapse at sintering or saturate the material (Figure 3a), and spheroidizing of molybdenum particles and re-crystallization through the liquid phase in the conditions of sintering the composite consisting of molybdenum and copper (Figure 3b). For comparison, the structure is shown in Figure 3c obtained from tested W-Ni powders. The structure of the materials was studied by means of scanning electron microscopy (SEM) of the secondary electrons. A variety of materials characterization including surface spectroscopy and X-ray based diffraction is used to assess the condition of the materials after processing with sintering-based techniques. The powder composite materials are so-called pseudo alloys, which provide promising high thermal conductivity properties, while displaying sub-unity sputter yields (see Section 4).

  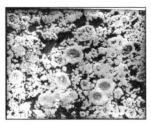

Fig. 3. From left to right, (a) the structure of the W-Cu-Ni-LaB6 pseudo alloy (x540), (b) the structure of the Cu-44%Mo – 1%LaB6 pseudo alloy (x2000), and (c) the structure of "irradiated" W-Cu-Ni pseudo alloy produced by class W-Ni powder (x400).

Observations made with secondary mass ion spectrometry (SIMS) on these materials found evidence of hydrogen and beryllium in anode components. Based on these results one can speculate that the hydrogen observed by SIMS after exposing the samples may be caused by that environment, in which the powders are manufactured, sintered, and additionally annealed. In regards to the beryllium observed on the anode surface after exposure to the xenon plasma, one may suppose two possible explanations, each of which requires additional verification. The construction may contain beryllium bronze; or the construction may contain Al_2O_3 or BeO based ceramics. Both cases may be the reason for enrichment of the surface samples by these elements during the heating phases.

For systems with the absence of the component interactions, the arc xenon plasma impact to the electrode materials does not cause a noticeable change of durability: for $MoCuLaB_6$: HV = 1600-1690 MPa; and for Cu- Al_2O_3: HV = 660 MPa through the whole height of the anode. In the tungsten and copper based composites, when presence of nickel exists, the mutual dissolution of the elements is increased (W is dissolved in Cu-Ni melt, for instance). At cooling, it may be accompanied by either forming non-equilibrium solid solution, or solidification; which is conformed by the increasing the firmness of the upper part of the anode (3380 MPa compared to 3020 MPa in its lower part). To provide more careful analysis, one should investigate the dependence of electro-conductive composites on heat resistance subject to arc discharges of powerful heat fluxes (up to 10^7 W/m²). Additional analyses typically conducted include the propagation of cracks, observed on the surface layer of the anode material and deep into the bulk. For that, the precise method of manufacturing is required for further insight on crack development and propagation. These analyses along with erosion material modeling (discussed in Section 4) are mainly used to dictate materials selection for electrode materials in EUV DPP sources.

2.3 EUV radiators, debris generation and debris mitigation systems

One particularly important "coupling" effect between the debris mitigation zone region and the collector optics region is the use of inert mitigation gases (e.g. Ar or He) that in turn are ionized by the expanding radiation field and thus generate low-temperature plasma near the collector mirror surface. This phenomenon is briefly discussed in Section 3. Each candidate radiator (e.g. Li, Sn or Xe or any combination) will result in a variety of irradiation-induced mechanisms at the collector mirror surface. For example, if one optimizes the EUV 13.5-nm light source for Li radiators, the energy, flux and mass distributions will be different compared to Sn. Both of these in turn are also different from the standpoint of contamination given that both are metallic impurities and Xe is an inert gas. The former will lead to deposition of material on the mirror surface. In the case of Xe, thermal deposition would be absent however the energetic Xe implantation on the mirror surface could lead to inert gas damage such as surface blistering and gas bubble production for large doses. Debris mitigation systems would have to be designed according to the radiator used.

3. EUV radiation-driven plasmas

As discussed earlier, Figure 1 shows the general configuration of a DPP system for EUV 13.5-nm light generation. Another "coupling" effect of the DMZ in the source system (e.g. from the electrode materials of the source through the DMZ to the collector mirror) is the fact that the intense EUV and UV radiation generated from the 13.5-nm radiators (e.g. Xe or Sn) can induce a secondary low-temperature plasma at the surface of the collector mirror by ionizing the protective gas used for debris mitigation such as argon or helium [Van der Velden et al, 2006, Van der Velden & Lorenz, 2008]. The characteristic plasma in this region is found to be of low temperature (e.g. 5-10 eV) and moderate densities (e.g. ~ 10^{16} cm^{-3}). The photoionization process can lead to fast electrons that induce a voltage difference the order of 70 V. In addition, due to the sheath region at the plasma-material interface between the plasma and the mirror the ionized gas particles (e.g. Ar$^+$ or He$^+$) can be accelerated up to about 50-60 eV. This energy in the case of Ar ions is relatively low and in the so-called sputter threshold regime for bombardment on candidate collector mirror material candidates. In addition, carbon contamination could also be accompanied by this plasma exposure. These candidate materials are typically thin (~20-60 nm) single layers of Ru, Rh or Pd, all of which reflect 13.5-nm light very efficiently. Only few studies have been conducted to elucidate how these low-energy ions may induce changes that can degrade the optical properties of the 13.5-nm collector mirrors. Van der Velden and Allain studied this effect in detail in the *in-situ* experimental facility known as IMPACT to determine the sputter threshold levels at similar energies [Allain et al, 2007]. In the work by van der Velden et al. the threshold sputtering of ruthenium mirror surface films were found to be in close agreement with theoretical models by Sigmund and Bohdansky. The sputter yields varied between 0.01-0.05 atoms/ion for energies about 50-100 eV and models were found to be within 10-15% of these values.

4. Irradiation-driven mechanisms on material surfaces

Before discussion of collector mirror geometry and configuration a brief background on irradiation-driven mechanism on material surfaces is in order. In DPP EUV devices electrodes at the source are exposed to short (10-20 nsec) high-intensity plasmas leading to a variety of erosion mechanisms. Erosion of the electrodes is dictated by the dynamics of the plasma pinch for configurations such as: dense plasma focus, Z-pinch and capillary. The transient discharge deposits 1-2 J/cm^2 per pulse on electrode surfaces. Large heat flux is deposited at corners and edges leading to enhanced erosion. Understanding of how particular materials respond to these conditions is part of rigorous design of DPP electrode systems. Erosion mechanisms can include: physical sputtering, current-induced macroscopic erosion, melt formation, droplet, and particulate ejection [Hassanein et al, 2008]. Erosion at the surface is also governed by the dynamics of how plasma can generate a vapor cloud leading to a self-shielding effect, which results in ultimate protection of the surface bombarded. Determining whether microscopic erosion mechanisms such as: physical sputtering or macroscopic mechanisms such as melt formation and droplet ejection the dominant material loss mechanism remains an open question in DPP electrode design. This is because such mechanisms are inherently dependent on the pinch dynamics and operation of the source. One important consequence of the extreme conditions electrode and collector optics surfaces are exposed is the existence of several irradiation-driven mechanisms that can lead to substantial materials mixing at the plasma-material interface. Bombarment-induced modification of materials can in principle lead to phase transition mechanisms that can substantially change the mechanical properties of the material accelerating degradation.

Conceptually, the phenomenon of bombardment-induced compositional changes is simplest when only athermal processes exist such as: preferential sputtering (PS) and collisional mixing (CM). Preferential sputtering occurs in most multi-component surfaces due to differences in binding energy and kinematic energy transfer to component atoms near the surface. Collisional mixing of elements in multi-component materials is induced by displacement cascades generated in the multi-component surface by bombarding particles/clusters and is described by diffusion-modified models accounting for irradiation damage. Irradiation can accelerate thermodynamic mechanisms such as Gibbsian adsorption or segregation (GA) leading to substantial changes near the surface with spatial scales of the order of the sputter depth (few monolayers). GA occurs due to thermally activated segregation of alloying elements to surfaces and interfaces reducing the free energy of the alloy system. Typically, GA will compete with PS and thus, in the absence of other mechanisms, the surface reaches a steady-state concentration approaching that of the bulk. However when other mechanisms are active, synergistic effects can once again alter the near-surface layer and complex compositions are achieved. These additional mechanisms include: radiation-enhanced diffusion (RED) due to the thermal motion of non-equilibrium point defects produced by bombarding particles near the surface, radiation-induced segregation (RIS), a result of point-defect fluxes, which at sufficiently high temperatures couples defects with a particular alloying element leading to compositional redistribution in irradiated alloys both in the bulk and near-surface regions. Figure 4 shows the temperature regime where these mechanisms are dominant. All of these mechanisms must be taken under account in the design of proposed advanced materials for the electrodes and the collector optics in addition to considering other bombardment-induced conditions (i.e., clusters, HCI, neutrals, redeposited particles, debris, etc...) that can be generated at the 13.5-nm light tool.

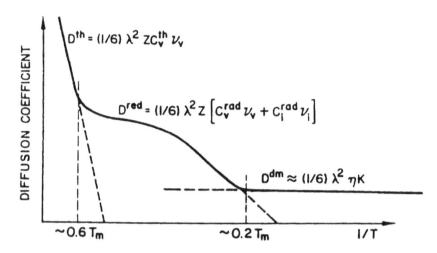

Fig. 4. Schematic plot of the relative importance and temperature dependence of displacement mixing, radiation-enhanced and thermally-activated mechanisms (e.g., Gibbsian segregation).

Modeling of physical sputtering is well known and the field quite mature, see for example work by W. Eckstein (Eckstein 1991) and W. Möller (Möller 1988). For energies above about 100 eV, binary collision approximation (BCA) codes are often used to estimate erosion from various material surfaces. The sputtering yield of 100% Cu from 1 keV Xe^+ bombardment coincides with the experimental result shown for Cu bombardment. Furthermore, example in Figures 5a and 5b, the sputtering from a W-Cu alloy is modeled. The advances in multi-scale and multi-component modeling provided by Monte Carlo damage codes such as TRIM-SP, TRIDYN and ITMC enables scoping studies of candidate materials and their surface response.

An additional mechanism currently missing in plasma-material interaction computational codes is the correlation of surface morphology with surface concentration. Ion-beam sputtering is known to induce morphology evolution on a surface and for multi-component material surfaces plausibly driven by composition-modulated mechanisms [Carter, 2001; Muñoz-Garcia et al., 2009]. Chason et al. have devised both theory and experiments to elucidate on surface patterning due to ion-beam sputtering [Chan & Chason, 2007]. A number of efforts also are attempting to enhance the ability to model ion-irradiation induced morphology and surface chemistry including work by Ghaly and Averback using molecular dynamics and by Heinig et al. using MD coupled KMC (kinetic Monte Carlo) approaches [Ghaly et al, 1999; Heinig et al., 2003]. In spite of these efforts there remains outstanding issues in ion-beam sputtering modification of materials such as the role of mass redistribution that can dominate over surface sputtering mechanisms [Aziz, 2006; Madi et al., 2011]. These developments have important ramifications to the EUV collector mirror operation given the complexity of energetic and thermal particle-surface coupling.

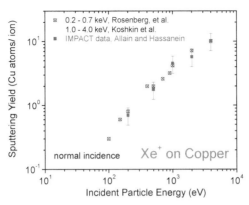

Fig. 5a. Sputtering yield of copper bombarded by singly-charged xenon at normal incidence in the IMPACT (Interaction of Materials with charged Particles And Components Testing) experiment at the Argonne National Laboratory.

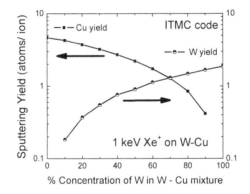

Fig. 5b. The Ion Transport in Materials and Compounds code developed at the Argonne National Laboratory calculates the partial sputtering yield of Cu and W from a W-Cu mixed material bombarded by 1 keV Xe ions at normal incidence. This system is used as a pseudo-alloy with properties able to withstand large heat fluxes in EUV source devices.

5. Collector mirrors for EUV lithography

The nature of the collector mirror damage is largely dictated by the configuration designed to optimize collection of the 13.5-nm light. Due to the refractive index in the X-ray and EUV range being less than unity, total external reflection is possible at angles that are large with respect to the mirror surface plane. If the geometry for collection of the light is such that the mirrors must collect light at more grazing incidence, than the configuration consists of collector mirrors with very thin single-layer coatings of candidate materials such as Ru, Pd or Rh. As discussed earlier the configuration in current EUV source technologies consist of either normal incidence mirrors or grazing incidence mirrors. The latter configuration must use a collection of multiple shell collectors designed to optimize collection of the 13.5-nm light. Media Lario, a lens manufacturer based in Italy, has optimized the multiple shell collector design in recent years.

5.1 Normal incidence mirrors

The normal incidence mirror configuration consists of a multi-layer mirror geometry exposed to 13.5-nm at normal incidence to the mirror surface. Due to the low reflectance fractions at normal incidence 10's of bilayers are stacked on top of each other to improve the reflectivity to the order of 50-60%. The mechanisms of radiation-induced damage depend on the mirror configuration as eluded above. In the case of the multi-layer mirror (MLM) the incident radiation is predominantly at near-normal incidence thus with the highest projected range into the material bulk. Intrinsic in the configuration of MLM collector systems is the inherent energy distribution of energetic particles that emanate from the LPP pinch plasma source. Although it is not a necessary requirement that MLM are used with LPP sources, the limited collection efficiency of grazing incidence mirrors motivate their use. However, in the context of irradiation damage from the nearby plasma MLM systems suffer the greatest losses in optical performance compared to GIM. The reason is two-fold. One the energy distribution from LPP sources tends to be dominant in the keV range of energies typically about 0.5-5-keV. Therefore there is immediate damage and ion-induced mixing at the MLM interfaces critical to the optimum reflectance of these mirrors. The use of Xe or Sn radiators also introduces a second challenge.

5.2 Grazing incidence mirrors

Grazing incidence mirrors are collector mirrors that reflect EUV light at angles that are predominantly inclined along the plane of the mirror surface. Since the collector mirror will have an inherent curvature the incident angle on the surface plane will have a variable incidence angle depending on the sector the light is collected. Furthermore, recent developments in grazing incidence mirror technology (e.g. Media Lario designs) have now optimized grazing incidence mirrors as shells with a hyperbolic, parabolic or ellipsoidal geometric curvature that optimizes the light collection. Typically the collector angle is about 5-25 degrees from the surface normal. In the grazing incidence mirror configuration there exists a number of issues in the context of irradiation-induced effects. For example the sputter efficiency of materials increases as the angle of incidence becomes more oblique. Therefore with this configuration there is the concern that the mirror could erode more rapidly. On the other hand, the implanted energetic debris is found closer to the surface, which could in some cases prove to be of benefit. The issue of incidence angle and its impact on both sputtering of the mirror material and the effect on EUV 13.5-nm reflectivity is discussed in later sections. Grazing incidence mirrors also entail only single layer materials in general. This is because the inherent light transport is via reflection and at grazing incidence typically a large fraction (> 60-70%) of the light can be reflected by materials such as: niobium, rhodium, ruthenium and palladium.

6. Irradiation modification of EUV optical properties

During a Sn-based LPP or DPP pinch, metal vapor will expand and reach nearby components including the collector mirror. Sn+ energies ranging from several hundred electron volts up to a few keV can be expected from Sn-based LPP or DPP source configurations and therefore constitute the energy range of interest for EUV collector mirror damage evolution. In the years between 2004 and 2007 Allain et al. conducted a series of pioneering experiments at Argonne National Laboratory. The work included a systematic *in-situ* characterization study in IMPACT of how candidate EUV mirror surfaces evolved under exposure to thermal and energetic Sn. Fig. 6 below depicts the various interactions relevant to the EUV 13.5-nm light source environment with candidate grazing incidence mirror materials: Ru, Pd or Rh. In this section studies on these materials and also candidate multi-layer mirror (MLM) materials are discussed with implications of ion-induced damage.

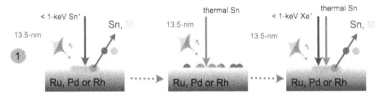

Fig. 6. Schematic of various interactions studied in IMPACT using Sn thermal and energetic particles while *in-situ* characterizing the evolving surface.

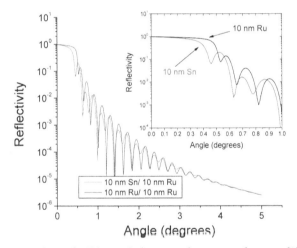

Fig. 7. X-ray reflectivity (8.043 keV X-rays) theoretical response for two different top surfaces: a 10-nm Sn surface on a 10-nm Ru underlayer and a 10-nm Ru surface on a 10-nm Ru underlayer, both with 0.5-nm rms roughness value (CXRO calculations).

The mirror reflectivity response at 13.5-nm light will be sensitive to the thickness of the deposited Sn layer. In addition, the reflectivity response may also be influenced by the structure of the material namely: evaporated porous structure, ion-induced densification phases and possible oxidation effects. All of these can be studied using XRR and in-band EUV reflectivity. When comparing for example a thin Sn layer to a thin Ru layer, theoretically, with enough Sn deposited, the extension of the critical edge will be reduced in the XRR response using CuKα X-rays.

Note the comparison made in Figure 7 showing CuKα (8.043 keV) X-ray reflectivity calculations using CXRO calculations for 10 nm Sn/Ru and 10 nm Ru layers with 0.5 nm rms roughness vs. incident grazing angle [Henke et al, 1993]. In the XRR vs. θ plot, the reflectivity suddenly decreases as θ^{-4} at angles above the critical angle, θ_c, which in this case it is equal to 0.45 degrees and 0.35 degrees for the 10-nm Ru and 10-nm Sn/Ru mirrors, respectively. The presence of the Sn layer effectively reduced the critical edge region and thus its reflectivity performance is reduced. This is because the momentum transfer, Q, is:

$$Q = \left(\frac{4\pi}{\lambda}\right)\sin\theta \qquad (1)$$

And the reflectivity response is related by:

$$Q_c^2 = 16\pi\rho_e \propto |R| \qquad (2)$$

This reflectivity response can also be assessed for the EUV spectral region (in-band 13.5-nm). Figure 8a shows CXRO calculations of the EUV in-band reflectivity response for same conditions in Figure 7. Note the reduction of the critical edge for the case of Sn deposition with a 10-nm Sn layer on top of a 10 nm Ru SLM.

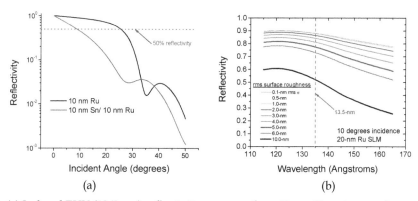

(a)   (b)

Fig. 8. (a) In-band EUV (13.5-nm) reflectivity response for a 10-nm Ru mirror and same mirror with a 10-nm Sn cap, and (b) theoretical calculations (CXRO) of in-band EUV reflectivity response versus incident angle at 13.5 nm (92 eV) for Ru and Sn surfaces.

Figure 8b shows the effect that surface roughness (e.g. morphology) can have on the absolute in-band (11-17 nm) EUV reflectivity from a 20-nm mirror Ru film surface. This is a great example of how both multi-component surface concentration (e.g. Sn particles in a Ru mirror surface) can couple with surface morphology evolution during deposition. Both a concentration of Sn and surface roughness can combine to decrease the reflectivity near 13.5-nm. The key question is what is the threshold for damage and can this be mitigated so that in steady-state a tolerable and minimal loss of reflectivity can be managed.

Figure 9 show AES data on a thin Ru-cap MLM before and after exposure to Sn vapor in IMPACT, respectively. In-situ metrology in IMPACT allows us to monitor in real time deposition of Sn on the mirror surface. EUV reflectivity from a MLM is near normal and thus the effect of a thin Sn layer must also be assessed as was done for the grazing incidence mirror data above. Figure 9b shows two major contaminants on the near surface (down to about 50-100 Å), oxygen and nitrogen.

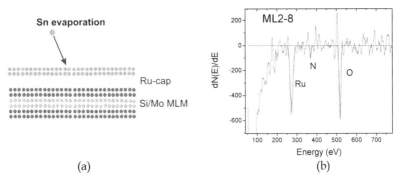

(a)   (b)

Fig. 9. (a) Schematic of Sn on MLM system. (b) Auger spectra of a thin Ru-cap MLM showing the presence of oxygen on the thin-film Ru cap. This MLM system can reflect up to about 69-72% of EUV light even in the presence of oxygen.

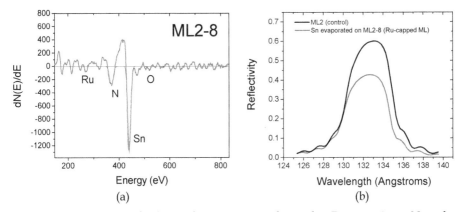

(a)   (b)

Fig. 10. (a) Auger spectra of a thin Sn layer evaporated on a thin Ru-cap mirror. Note the presence of nitrogen as opposed to oxygen and the strong Sn peak and (b) In-band EUV reflectivity data taken at NIST-SURF facility. Note the noticeable effect on the reflectivity response for the ML2-8 sample.

Oxygen is always found on the surface in the presence of ruthenium due to its high oxygen affinity. When a thin layer of Sn is deposited as shown in Figure 10a, the major contaminant is nitrogen and not oxygen. This is due to tin's high affinity for nitrogen compared to oxygen. Figure 10b shows the effect of an evaporated Sn layer on EUV mirror reflectivity. The EUV in-band reflectivity was measured at the NIST-SURF facility at near-normal incidence. The reduction from about 60% in-band EUV reflectivity to about 40% is consistent with deposition of about a 40-50 Å Sn thin layer. This has been corroborated by calculations on a thick Ru surface layer at near-normal incidence, giving a thickness comparable to about 35 Å.

6.1 Effect of surface roughness on 13.5-nm reflectivity

The effect of the surface evolution (e.g. concentration and morphology) on 13.5-nm reflectivity is a key factor in determining the lifetime of the collector mirror during operation of the high-intensity EUV lamp. A number of *in-situ* characterization studies are conducted to study the evolution of the surface structure, concentration and morphology under relevant EUV light generation conditions. Single-effect studies are presented in this section to illustrate and differentiate effects from the expanding thermal Sn plume and the energetic Sn particles that emanate from the high-density pinch Sn plasma region.

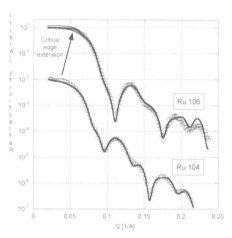

Fig. 11. Fits to Ru 106 and Ru 104. The Ru 104 data set and fit have been shifted downward by a factor of 100 for clarity. The electron density depth profiles from these fits are shown in Figure 12.

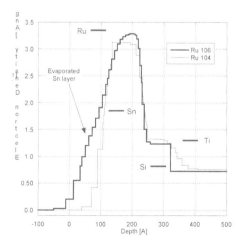

Fig. 12. Electron density profiles for Ru 106 and Ru 104. The presence of a rough Sn layer at the air-film interface of Ru 106 is clear. The bulk density values are shown as horizontal lines.

Fits to Ru 106 (with evaporated Sn layer) and Ru 104 (identical to Ru 106, but without Sn layer) are shown in Figure 11. The electron density depth profiles obtained from these fits are shown in Figure 12. First, the electron density values for Ru 106 and Ru 104 are consistent with the known bulk values. The presence of the Sn layer on Ru 106 is clear. In fact the point at which the profiles for Ru 106 and Ru 104 diverge (near the air-film interface) corresponds to the bulk Sn electron density value. Thus, the Ru 106 data set is consistent with a Sn over-layer approximately 60 Å thick. The air-Sn layer interface is not well defined as determined by the fit of the XRR data. The extension of the critical edge for Ru 106 is evident, an effect due to the Sn layer increasing the total electron inventory of the metal over-layer.

The evaporated Sn layer on this sample is either very rough, has significant internal porosity, or has intermixed with the Ru layer to a large extent. Surface roughness values above 5-nm rms would need to exist to lead to any significant decrease on in-band EUV reflectivity (as shown earlier in Fig. 8b). Significant intermixing is very possible during the low-energy room temperature evaporation. It is possible Sn does not wet Ru adequately and this could lead to a poor surface topography and a rough interface. The blurry Ti-Si interface for the Ru 104 sample probably is not a real effect, but a consequence of an incomplete fit.

The effect of the thin-film Sn layer on in-band (13.5-nm) EUV reflectivity is shown in Figure 13. Measurements were conducted at the NIST-SURF facility. The figure shows two primary cases. One is Ru-104, a virgin 10-nm Ru sample. Both XRR and QCM-DCU (quartz crystal microbalance dual-crystal unit) measurements of this particular batch of Ru SLM measured a Ru film thickness of about 100 Å [Allain et al, 2007]. The EUV in-band 13.5-nm reflectivity data fitted with CXRO calculations is yet a third indication of the Ru thin-film thickness, thus effectively calibrating the QCM-DCU data in *in-situ* characterization. The EUV reflectivity results show that the Ru thin-film thickness is about 90 Å fitting with the CXRO calculations. The sample covered with Sn (Ru-106) is fitted with CXRO calculations using a 2.1-nm Sn surface layer at 20-degrees incidence. This correlates well with estimates from Sn fluences measured in IMPACT giving about a 30-40 Å Sn thin-film layer.

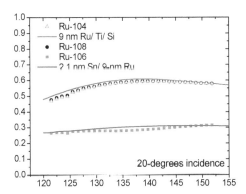

Fig. 13. Two virgin samples, Ru-104 and Ru-108 are shown with their reflectivity response in the EUV in-band 13.5-nm spectral range at 20-degrees with respect to the mirror surface. The reflectivity response of the Sn-covered mirror is also shown.

Fig. 14. SEM image of Rh-313 exposed to similar conditions as sample Ru-106. Therefore Sn coverage is equivalent to about a 2-nm thickness of Sn atoms.

The EUV reflectivity mirror response measured in-situ is correlated to ex-situ surface morphology data using SEM and EDX for electron-based microscopy. Fig. 14 shows SEM data for the case of Rh-313 exposed to 50 nA of Sn evaporation for 15-minutes. The surface morphology is characterized by surface structures that vary in lateral size from 10-100 nm. Observations from BES (backscattering electron spectroscopy) data suggested that the lighter imaged structures correspond to Sn, while darker regions corresponded to Rh. This led to the conclusion that the surface structures are islands of Sn that have coalesced during deposition. The formation of these two-dimensional nanostructures could be associated with diffusion-mediated aggregation of deposited Sn atoms. This is partly due to deposition of tin driving the morphology and structure of the Sn film deposited on the SLM surface far from equilibrium conditions. When one incorporates the kinetic effect of energetic implanted Sn, the net energy available is increased dramatically. This point is further investigated in later sections. The formation and growth of nano-scale tin islands during exposure is a competition between kinetics and thermodynamic equilibrium of deposited Sn atoms on the surface of either of the noble metal used (e.g. Ru or Rh).

The results from a set of thin Ru films exposed to energetic Sn ions are shown in Figure 15. The SLD profiles exhibit the effect of sputter erosion caused by the Sn-ion bombardment. Although the fluence of Ru102 and Ru105 differed by a factor of approximately 20, the profiles are similar. This is probably the effect of greater sputter efficiency for the low fluence Ru 105 case where the ion irradiation angle was 45° instead of normal incidence.

6.2 Effect of fast and thermal particles on MLM reflectivity at 13.5-nm

For MLM systems, Xe^+-bombardment studies in IMPACT demonstrated that the main failure mechanisms were: 1) ion-induced mixing at the interfaces along with significant sputtering of cap material (i.e., Ru) and 2) synergy of energy (1-keV) and high mirror temperature (200° C) leading to mirror reflectivity degradation [Allain et al., 2006]. Therefore, from the point of view of ion-induced damage, MLM systems compared to SLM systems are most susceptible to early failure rates if fast ion and neutral energies are maintained at the 1 keV level or more.

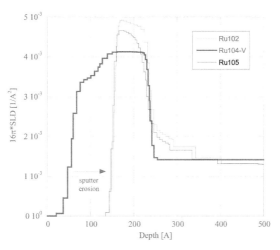

Fig. 15. Electron density depth profiles (ordinate is equal to 16πSLD, where SLD=$r_e\rho_e$, and r_e and ρ_e are the classical electron radius and electron number density, respectively). The overall film thickness for Ru102 and Ru105 has been reduced by sputtering.

Kinematically, Xe and Sn behave similarly, since their mass is very close. However, there is a fundamental difference: unlike Xe (which is inert), Sn can be incorporated into the mirror structure and easily build up on the target. Sn accumulation would be exacerbated if any type of chemical bonding or new phase is formed. The accumulation of Sn is limited during Sn bombardment due to self-sputtering; therefore a steady-state Sn content in the sample is reached. In addition, the overall ion-induced sputtering of the mirror is reduced, since ion-induced sputtering is now shared between the mirror material (i.e., Ru, Rh or Pd) and the previously implanted Sn. Results from Monte Carlo modeling of Sn implantation have shown these trends, and they were later verified by experimental measurements [Allain et al., 2006]. Tests therefore conducted with Xe+ served as an appropriate surrogate for Sn irradiation. Furthermore, since some EUV light sources could in principle use Xe as a 13.5-nm radiatior, these tests were also directly relevant. One particular interesting effect of inert ions such as Xe is that they implant at the near surface and could, if enough vacancy-induced voids are created, lead to Xe bubble accumulation. The work by Allain et al. in fact now has indicated that for a given Xe fluence threshold at 1-keV the stability of small nm-sized bubbles can be created at the near surface of MLM Si/Mo systems. This was indicated by use of XRR tests showing Porod-like scattering of small-angle X-ray scattering experiments.

6.3 Effect of fast and thermal Sn particles on single-layer reflectivity at 13.5-nm
6.3.1 Thermal Sn
Operation of Sn-based EUV lithography DPP sources exposes the collector mirror to two types of Sn contamination: thermal deposition of Sn vapor and bombardment of Sn ions from the expanded plasma. Even with the implementation of debris mitigation mechanisms, some contamination will reach the collector mirror. In the *in-situ* expeirments presented here, both sources of Sn (i.e., energetic and thermal) can be studied on small mirror samples. An electron beam evaporator loaded with Sn supplies the

thermal flux. The energetic Sn flux comes from a focused Sn-ion source. Integration of an *in-situ* EUV reflectometer allows monitoring of the reflectivity in real time as the mirror is exposed to Sn.

EUV reflectivity measurements were monitored as the Sn layer was deposited. Results from these Sn exposures are shown in Figure 16. The lower axis corresponds to the Sn fluence and the thickness of the deposited Sn layer (calculated assuming that the film density is equal to the Sn bulk density) in the upper axis. For the case of the Rh sample (Rh-211) the Sn layer thickness is calculated based on fits with the reflectivity code and absolute at-wavelength 13.5-nm data from NIST. For a 15 nA current on an ECN4 evaporator for 2 minutes, sample Rh-213 was used as calibration sample with similar conditions to Rh-211. The sputter rate measured was 0.048 nm/sec or 2.9 nm/min. For Rh-211, the current level used was 5 nA for 34 minutes. This results in a deposition rate of 0.125 nm/min (2.9 divided by a factor of 3 and 7.75) and multiplied by 34 minutes results in a thickness of about 4.25 nm. Ex-situ XRF measurements resulted in an equivalent Sn thickness of 3.14 nm. The result appears consistent between the independent XRF measurement and the known deposition rate measured in the *in-situ* experiments in IMPACT. However, there are two observations with this result when one examines Fig 16 more carefully. One is the fact that the surface atomic fraction never reaches 100% of Sn atoms to Rh for Rh-211. Since LEISS is sensitive only to the first monolayer and the thickness measured is about 4-nm, one would expect LEISS to only scatter from Sn atoms at the surface. The LEISS data shows that instead an equilibrium concentration is reached near 70%. The second issue pertains to the in-situ relative reflectivity measured. For levels of 4-nm Sn deposition one would expect the relative reflectivity loss is of order 40-50% losses. However, the measurements show that losses in reflectivity are only about 20-30%. This is in direct contradiction to theoretical results of a Sn 4-nm layer on Rh. To investigate this further, a different mirror substrate (Pd) is used with similar Sn exposure conditions.

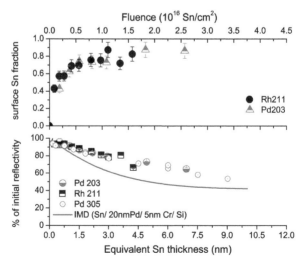

Fig. 16. Evolution of the EUV reflectivity for a Rh mirror as a Sn layer is deposited on the surface compared to deposition on a Pd mirror.

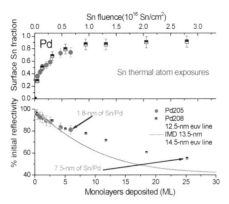

Fig. 17. Evolution of the EUV reflectivity for two Pd mirrors as a Sn layer is deposited on the surface.

Sn thermal deposition on Pd mirrors show similar behavior using about 10 nA of Sn thermal current. This corresponds to a deposition rate of 0.25 nm/min on Pd-203 (2.9 divided by a factor of 1.5 and 7.75) and for a 28 minute exposure a Sn thickness of about 7.5 nm. The XRF measurements resulted in a 6.46 nm equivalent Sn thickness, in reasonable agreement with IMPACT deposition rate measurements. In Figure 16 the relative reflectivity loss is about 35% for Pd-203, much lower than theoretically predicted for deposition of a 7.5-nm Sn layer on Pd.

For the cases of Pd 205 and 208, the deposition rate is 4-5 times less than for Pd-203. This is based on the time of equilibration of the Sn surface atomic fraction measured by LEISS of Pd 205 and 208 compared to Pd 203. Therefore, the deposition rate for Pd 205 and Pd 208 is about 0.0625 nm/min. For Pd-208 and 120 minute exposure the Sn thickness is 7.5 nm and for Pd 205, 28-minute exposure, 1.8 nm. The relative reflectivity losses are 20% and 45% for Pd-208 and Pd-205, respectively as shown in Figure 17. The surface atomic fraction of Pd-208 reaches 85-90% after close to 10^{16} Sn/cm² fluence. Before this time, for fluences below 0.6-0.7 x 10^{16} Sn/cm² the surface Sn atomic fraction reaches levels of about 70% for Pd-205 and Pd-208 consistent with results for Pd-203. So for exposures below Sn fluences of 10^{16} Sn/cm², the relative reflectivity losses are below about 30%. The main difference between Pd-203 and Pd-205, is that for the same exposure time (28 min.), Pd-203 has a "thicker" equivalent Sn layer compared to Pd-205 based on the deposition rate measured. This is an important result in that, although for the fluence exposure one should get "thick" Sn layers, the results from low-energy ion scattering shows otherwise. That is, LEISS is sensitive to the first or second monolayer and the data shows that even in the cases of Pd-203 and Pd-208 about 10-15% of the scattered ions detected, scatter from Pd atoms. Moreover, for lower fluences, scattering from mirror atoms (Pd or Rh) can be as large as 30%. More importantly, the surface Sn fraction seems to reach an equilibrium until the fluence is increased further.

These results imply that Sn is coalescing into nm-scale islands on the substrate surface for Sn exposures below about 10^{16} Sn/cm². Surface morphology examination was conducted with scanning electron microscopy (SEM) as a function of the Sn thermal fluence. The results were very important in that it proved that indeed the lower reflectivity loss is attributed to Sn island coalescence.

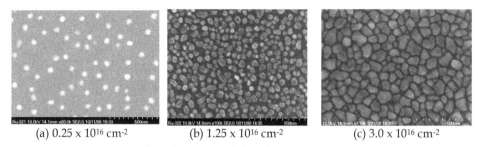

(a) 0.25×10^{16} cm^{-2} (b) 1.25×10^{16} cm^{-2} (c) 3.0×10^{16} cm^{-2}

Fig. 18. SEM micrographs of Sn-deposited Pd thin-film mirrors as a function of the Sn fluence.

6.3.2 Sn ions

EUVL plasma-based Sn sources expose mirrors to both thermal and energetic particles as discussed earlier. In this section we investigate the EUV reflectivity response of grazing incidence mirrors to exposure of Sn ions. The goal of this investigation was to identify failure mechanisms on the performance of Ru mirrors under EUVL source-relevant conditions. Furthermore, these experiments were also designed to elucidate the behavior of energetic Sn particles against results of thermal Sn exposure presented in the previous section. In addition to thermal Sn deposition on the collector mirror in a EUVL source device, the mirror is also subjected to energetic fast-ion and neutral bombardment from expanded plasma that gets through the debris mitigation barrier. The study of this problem is critical to assess the severity of damage induced by fast ion/neutral bombardment on EUV collector mirrors. Ion bombardment induces damage to EUV mirrors with at least three mechanisms: 1) erosion of the mirror material by physical sputtering, 2) modification of surface roughness, and 3) accumulation of implanted material inside the mirror.

These three phenomena have been extensively explored in IMPACT for the case of Xe+ bombardment, both for single-layer and multilayer EUV mirrors (Nieto et al, 2006). For this case, the first mechanism, erosion of the mirror, was determined to be the limiting factor for mirror lifetime. Surface roughness changes induced by ion bombardment in those cases were not large enough to affect the reflectivity in a significant manner. This was consistent with findings of irradiated thin-film surfaces of mirrors fabricated with magnetron sputtering. Typically these films consist of large grain boundary density, and thus surface corrugated structures from ion-beam bombardment are minimized. In regards to accumulation, it was observed that large Xe fluences ($>10^{17}$ Xe$^+$/cm^2) delivered over a short period of time caused blistering of the mirror most likely due to the formation of bubbles. Xe fuel accumulation in the mirror layer is not regarded as an issue for sources operating with Xe$^+$ at low EUV power operation. Under high-power HVM (high-volume manufacturing) level operation, with Xe as the EUV radiator, it's unclear how large dose exposures might scale. Suffice to say that if the Xe flux is not controlled and maintained at tolerable levels, significant damage to the grazing incidence mirror is likely, mostly from ion-induced sputtering (Nieto et al, 2006).

Two experiments were performed by exposing two Ru mirrors to 1.3 keV Sn beams with a current of 40 -50 nA. The beams were rastered over a 0.25 – 0.3 cm^2 area, giving a net Sn ion flux of ~ 10^{12} ions cm^{-2} s^{-1}. The mirrors were exposed to this Sn beam for three hours (~10^4 sec), giving a total fluence of 10^{16} ions cm^{-2}. Sample ANL-H was manufactured by Philips, and Ru-208 was manufactured by OFM-APS at ANL. Sample ANL-H was bombarded at 60° incidence, while Ru-208 was bombarded at normal incidence. The results of the exposures are presented in Figure 19a and Figure 19b, which show both the Sn surface concentration (upper panels) and relative EUV reflectivity (lower panels).

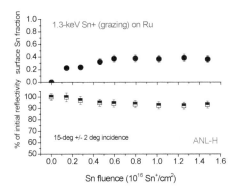

Fig. 19a. Evolution of the surface concentration and the EUV reflectivity of a Ru mirror exposed to a 1.3 keV Sn beam incident at grazing incidence (60°).

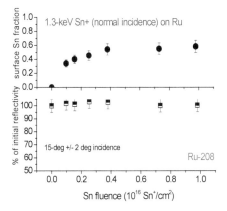

Fig. 19b. Evolution of the surface concentration and the EUV reflectivity of a Ru mirror exposed to a 1.3 keV Sn beam incident at normal incidence (0°).

There are significant differences between the two exposed samples. Regarding the Sn content in the surface, it can be seen that the sample bombarded at grazing incidence (ANL-H) reaches an equilibrium Sn content of 40%, while the sample bombarded at normal incidence has a steady-state Sn surface fraction of 60%. The increase can be explained by an increase of Sn self-sputtering yield. The Sn atomic fraction y_{Sn} on the sample as a function of time is given by:

$$\frac{dy_{Sn}}{dt} = \frac{\Gamma_{sn}}{n_T}\left(1 - y_{Sn}Y_{self\ sp}\right) \qquad (3)$$

Equation 3 represents the balance between the implantation and the sputtering flux. The implantation flux is constant, but the sputtered flux is actually a function of the Sn content in the sample, so it gets weighted by the atomic fraction of Sn in the target y_{Sn}. At equilibrium, the time derivative is zero and that condition relates the equilibrium Sn fraction $y_{Sn,eq}$ and the self sputtering yield of Sn, $Y_{self\ sp}$:

$$y_{Sn,eq} = \frac{1}{Y_{self\ sp}} \tag{4}$$

For 60° incidence, the equilibrium fraction is 0.4, which corresponds to a self-sputtering yield of 2.5. For the normal incidence bombardment, the Sn self sputtering yield corresponding to the 0.65 Sn equilibrium atomic fraction is 1.5. These numbers are very close to the ones reported in the literature for Sn self-sputtering. Therefore, this is yet another independent verification of the in-situ EUV reflectivity measurements in IMPACT.

The other interesting observations from [Allain et al, 2007b] and [Allain et al., 2010] relates to the behavior of the EUV reflectivity as Sn is implanted. The effect of implanted Sn is not as drastic as for the case of deposited Sn on the surface, since the change in reflectivity is very small. For the sample irradiated at normal incidence, the reflectivity does not drop at all during the irradiation over a fluence of 10^{16} Sn+/cm^2. For the sample exposed to the beam at 60° incidence, a drop of < 10% in reflectivity is observed. By comparing the fluence scales for figures 18 and 19, it can be seen that the deposited Sn produces a more pronounced drop on reflectivity (15%), a drop at least 3 times larger than the one observed for the samples with implanted Sn. The case for the grazing incidence irradiation produces a larger drop in reflectivity that the normal incidence case, since in the limit of completely grazing incidence (90°), the implantation and thermal deposition cases are basically the same, since there is no penetration into the target.

6.3.3 Sn or Xe ions combined with thermal Sn

To examine the effects of exposure to a more realistic environment in a EUV light tool with both energetic and thermal particles exposing the collector mirror surface, experiments with thermal Sn and energetic Xe+ were conducted. For these experiments, three samples— Rh 318, Rh 319, and Rh 320 —were each irradiated with a 1 keV ion beam (Xe+) and exposed to an evaporator (Sn) simultaneously, with a total exposure time of 36 minutes. The target energetic Xe+ fluences increased by one order of magnitude with each successive sample, beginning at 4.5×10^{15} Xe/cm^2, while target thermal Sn fluences remained constant at 4.5×10^{16} Sn/cm^2. Two control samples, Rh 321 and Rh 323, were used to compare the effects on reflectivity. Rh 321 was exposed to thermal Sn evaporator with a target Sn fluence of 4.5×10^{16} Sn/cm^2 for 36 minutes with no irradiation and Rh 323 was irradiated with an ion beam (Sn+) at 1.3 keV for 88 mins at a fluence of 1.03×10^{14} Sn+/cm^2 with no thermal Sn deposition.

Figure 20 shows both relative percent EUV reflectivity and Sn surface fraction versus thermal Sn fluence. A direct correlation between reflectivity loss and surface fraction of Sn is observed. Rh 318 and Rh 319 are fully covered with Sn after 3 minutes of exposure and their relative reflectivity decreased by 41.6% and 48.5%, respectively, after 36 minutes. While reflectivity of Rh 318 and Rh 319 decreased as the experiment progressed, Rh 320 had a local maximum at approximately 2.21×10^{16} Sn cm^{-2} where reflectivity increased to 94.7%. The corresponding Xe+ fluence, 2.25×10^{16} Xe+ cm^{-2}, exceeds the final fluences for the other two samples. This suggests that Rh 320 reached a threshold—too high for the other samples—

where the surface changed such that reflectivity could reach a maximum. The control sample, Rh 321, behaved extremely similar to Rh 318 and Rh 319 in both the atomic fraction of Sn and relative reflectivity loss. The relative reflectivity of Rh 323 fluctuated with increasing fluence but was found to only decrease 1.4% at the highest fluence, 5.8×10^{15} Sn^+/cm^2. Figure 5 does not represent the fluence corresponding to the reflectivity of Rh 323 because there was no thermal Sn fluence on the sample. It was plotted purely to show the affects of energetic Sn fluence on reflectivity. The surface fraction of Sn was lowest for this sample when compared to the other Rh samples, which was expected, with the surface fraction of Sn approaching equilibrium at 0.484. This further confirms the direct correlation between reflectivity loss and surface fraction of Sn discussed earlier.

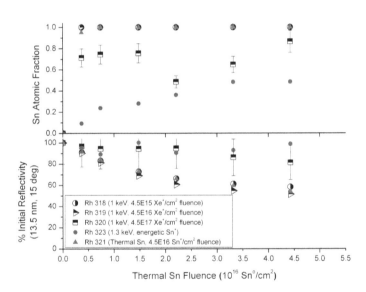

Fig. 20. LEISS data showing the surface Sn fraction versus thermal Sn fluence (top) and 13.5nm EUV reflectivity measurements versus thermal Sn fluence (bottom). Rh 321 had thermal Sn deposition only at a fluence of 4.5E16 Sn^0/cm^2. The fluence of Rh 323 shown, for both cases, is meant for correlation purposes only since there is no thermal Sn fluence on the sample.

	10x10 µm	10x10 µm	10x10 µm
Sample	Rh 318	Rh 319	Rh 320
Area Ra (nm)	18.5	14.8	2.00
Area RMS (nm)	22.4	19.4	3.20
Average height (nm)	60.9	16.7	1.78
Max. height (nm)	137	100	26.4

Table 1. Two-dimensional (10x10 µm) atomic force microscope (AFM) images and roughness values calculated with the AFM computer analysis program.

	5x5 µm	5x5 µm
Sample	Rh 321	Rh 323
Area RMS (nm)	14.12	0.11
Max. height (nm)	89.00	2.05
Feature Area (um^2)	22.98	n/a
Feature Coverage (%)	91.9%	n/a

Table 2. Two-dimensional (5x5 µm) atomic force microscope (AFM) images and roughness values calculated with the AFM computer analysis program for Rh thermal Sn only (left) and energetic Sn only (right) samples.

The AFM investigated the morphology of the samples. Table 1 and 2 illustrate the results. As the fluence of the samples is increased, it is found that the height, the roughness and the general size of the features decrease significantly. This is likely due to the increase in sputtering of Sn caused by the higher Xe^+ fluence. Rh 318 has the largest roughness and height values, at 22.4 nm and 137 nm respectively, but has the lowest fluence of the sample set. Rh 320 on the other hand has the smallest values for roughness and height, 3.20 nm and 26.4 nm, but the largest fluence of the set. This shows a direct correlation between the morphology of the samples and the fluence of Xe^+.

There is also a correlation between the morphology and the resulting reflectivity. For these samples, the lower roughness value (and height value) corresponds to the highest reflectivity. This is seen with Rh 320, where the reflectivity drop is 18.7% and the RMS value for the roughness is 3.20 nm, the lowest value for each in the sample set. For Rh 318 and Rh 319, the roughness values are 22.4 nm and 19.4 nm and the height values are 60.9 nm and 16.7 nm with corresponding reflectivity losses of 41.7% and 48.5%, respectively.

The difference in surface morphology for thermal Sn only and energetic Sn only is clearly illustrated in table 2. The roughness for thermal Sn deposition only, Rh 321, was found to be 14.12 nm, almost identical to Rh 319, 14.8 nm, and very close to Rh 318, 18.5 nm. This closeness is surface roughness, as well as similarities in reflectivity loss and Sn atomic fraction, elucidates the correlation between surface morphology and resulting reflectivity. This is further cemented by comparing the reflectivity loss of Rh 323 with its surface morphology. Rh 323 had a surface roughness of only 0.11 nm and maximum feature height of 2.05 nm with its largest drop in reflectivity being only 11%.

7. Conclusion

In conclusion, the success of EUV lithography as a high-volume manufacturing patterning tool remains elusive although great progress has been made in the past half decade. One main challenge is the plasma-facing components (e.g. electrodes, collector mirrors and debris mitigation shields) lifetime that ultimately impact the EUV power available for exposure.

8. Acknowledgements

We acknowledge the Intel Corporation, Dr. Melissa Shell, Dr. Bryan Rice (currently as an Intel at Sematech SUNY Albany), and Dr. Robert Bristol. We thank our strong collaborations with numerous groups including the groups of: Dr. Charlie Tarrio at NIST, Prof. Brent J. Heuser at University of Illinois, Dr. Peter Zink formerly from Philips Research Labs, Dr. Al Macrander of the Optics Laboratory at Argonne National Laboratory, Dr. Sasa Bajt formerly of Lawrence Livermore National Laboratory, and Dr. Vivek Bakshi formerly of Sematech.

9. References

Allain, J. P.; Hassanein, A. et al, "Effect of charged-particle bombardment on collector mirror reflectivity in EUV lithography devices", Proc. SPIE Int. Soc. Opt. Eng. 6151 (2006) 3

Allain, J. P.; Nieto, M.; Hendicks, M.; Harilal, S. S.; Hassanein, A (2007). *Debris-and radiation-induced damage effects on EUV nanolithography source collector mirror optics performance,* SPIE Proceedings, 6586(2007):22, doi: 10.1117/12.723692

Allain, J.P., Nieto, M., Hendricks, M.R., Plotkin, P., Harilal, S.S, & Hassanein, A (2007). IMPACT: A facility to study the interaction of low-energy intense particle beams with dynamic heterogeneous surfaces. Rev. Sci. Instrum. 78, 113105 (2007), ISSN 0034-6748

Allain, J.P.; Nieto-Perez, M.; Hendricks, M.R.; Zink, P.; Metzmacher, C.; & Bergmann, K. (2010). *Energetic Sn+ irradiation effects on ruthenium mirror specular reflectivity at 13.5-nm*. Applied Physics A, 100, 1, pp. 231-237, (July 2010), ISSN 0947-8396

Allain, J.P.; Nieto, M.; & Hassanein, A. (2008). *Specular reflectivity of 13.5-nm light from Sn islands deposited on grazing incidence mirror surfaces. Applied Physics A*, 91, 1, pp. 13-16, (April 2008), ISSN 0947-8396

Aziz, M. J. (2006). *Nanoscale Morphology Control Using Ion Beams*, Proceeding in Ion Beam Science: Solved and Unsolved Problems, Matematisk-fysiske Meddelelser 52, Sigmund, P. (editor), ISBN: 87-7304-330-3

Bakshi, V. Editor (2006). *EUV Sources for Lithography*, SPIE, Bellingham, WA, ISBN 0819458457

Bakshi, V. Editor (2009). *EUV Lithography*, SPIE and John Wiley & Sons, Hoboken, New Jersey, ISBN 978081946949

Banine, V. & Moors R. (2004). *Plasma sources for EUV lithography exposure tools*, J. Phys. D: Appl. Phys. 37 3207

Banine, V.; Koshelev, K.N.; Swinkels G.H.P.M. (2011) *Physical processes in EUV sources for microlithography*, J. Phys. D: Appl. Phys. 44 253001

Benschop, J.; Banine, B.; Lok, S.; & Loopstra, E. (2008). *Extreme ultraviolet lithography: Status and prospects*, J. Vac. Sci. Technol. B 26, 2204, ISSN 1520-8567

Campos, D.; Harilal, S. S.; & Hassanein, A. (2010). *The effect of laser wavelength on emission and particle dynamics of Sn plasma*, J. Appl. Phys. 108, 113305, ISSN 1089-7550

Carter, G., (2001). *The physics and applications of ion beam erosion*. Journal of physics. D, Applied physics 34.3, doi: 10.1088/0022-3727/34/3/201

Chan, W; Chason E (2007). *Making waves: Kinetic processes controlling surface evolution during low energy ion sputtering*, J. Appl. Phys. 101, 121301, DOI:10.1063/1.2749198

Eckstein, W. *Computer simulation of ion-solid interactions*, Springer Series in Materials Science, Vol. 10, Springer, Berlin, 1991, ISBN 3-540-190570-0

Fahy, K.; O'Reilly, F.; Scally, E.; & Sheridan, P. (2010). *Robust liquid metal collector mirror for EUV and soft x-ray plasma sources*, Proceedings Vol. 7802 in Advances in X-Ray/EUV Optics and Components V, Goto, S.; Khounsary, A. M.; & Morawe, C., Editors (2010). 78020K, 27 August 2010, Proc. SPIE 7802, 78020K (2010); doi:10.1117/12.860747

Ghaly, M.; Nordlund, N.; Averback, R.S. (1999). *Molecular dynamics investigations of surface damage produced by kiloelectronvolt self-bombardment of solids*, Philosophical Magazine A, vol. 79, Iss. 4, p.795-820, doi: 10.1080/01418619908210332

Harilal, S. S.; O'Shay, B.; Tillack, M. S.; Tao, Y.; Paguio, R.; Nikroo, A.; & Back, C. A. (2006). *Spectral control of emissions from tin doped targets for extreme ultraviolet lithography*, J. Phys. D: Appl. Phys. 39 484

Hassanein, A.; Sizyuk, V.; and Sizyuk, T. (2008). *Multidimensional simulation and optimization of hybrid laser and discharge plasma devices for EUV lithography*, Proc. SPIE 6921, 692113-1-15, DOI:10.1117/12.771218

Heinig, K.H.; Muller, T; Schmidt, B; Strobel, M.; Moller, W. (2003). *Interfaces under ion irradiation: growth and taming of nanostructures*, Appl. Phys. A, vol. 77, iss. 1, pp. 17-25, ISSN: 0947-8396

Henke, B. L.; Gullikson, E. M.; Davis, J.C. (1993). *X-ray interactions: photoabsorption, scattering, transmission, and reflection at E=50-30000 eV, Z=1-92*, Atomic Data and Nuclear Data Tables, Vol. 54, 181-342, Available from: http://henke.lbl.gov/optical_constants

Jurczyk, B. E.; Vargas-Lopez, E.; Neumann, M. N.; & Ruzic, D. N. (2005). *Illinois debris-mitigation EUV applications laboratory*, Microelectronic Engineering, Volume 77, Issue 2, February 2005, Pages 103-109, ISSN 0167-9317

Madi, C. S.; Anzenberg, E.; Ludwig, K.; Aziz M.J. (2011). *Mass Redistribution Causes the Structural Richness of Ion-Irradiated Surfaces*, Physical review letters 106.6, 66101.

Möller, W.; Eckstein, W.; Biersack, J.P. (1988). *Tridyn-binary collision simulation of atomic collisions and dynamic composition changes in solids*, Computer Phys. Comm. 51, No. 3, 355-368.

Muñoz-Garcia, J; Vazquez, L.; Cuerno, R.; Sanchez-Garcia, J.; Castro, M.; Gago, R. (2009). *Self-Organized Surface Nanopatterning by Ion Beam Sputtering*, in Toward Functional Nanomaterials: Lecture Notes in Nanoscale Science and Technology, Whang, Z (Ed.), Springer-Verlag, pp. 323 – 398, ISBN 978-0-387-77716-0

Nieto, M.; Allain, J.P.; Titov, V.; Hendricks, Matthew R.; Hassanein, A.; Rokusek, D.; Chrobak, C.; Tarrio, Charles; Barad, Y.; Grantham, S.; Lucatorto, T.B.; Rice, B. (2006). *Effect of xenon bombardment on ruthenium-coated grazing incidence collector mirror lifetime for extreme ultraviolet lithography*, J. of Appl. Phys., Vol. 100 Issue 5, p053510, ISSN: 00218979

O'Connor, A.; Dunne, P.; Morris, O.; O'Reilly, F.; O'Sullivan, G.; & Sokell, E. (2009). *Investigation of ions emitted from a tin fuelled laser produced plasma source*, J. Phys.: Conf. Ser. 163 012116

Tarrio, C. & Grantham S. (2005). *Synchrotron beamline for extreme-ultraviolet multilayer mirror endurance testing*, Rev. Sci. Instrum. 76, 056101, ISSN 0034-6748

Thompson, K. C.; Antonsen, E. L.; Hendricks, M. R.; Jurczyk, B. E.; Williams, M.; & Ruzic, D. N. (2006). *Experimental test chamber design for optics exposure testing and debris characterization of a xenon discharge produced plasma source for extreme ultraviolet lithography*, Microelectronic Engineering, Vol. 83, Iss. 3, pp. 476-484, (March 2006), ISSN 0167-9317

Van der Velden, M; Lorenz, M. (2008). *Radiation Generated Plasmas: a challenge in modern lithography*, Technische Universeit Eindohoven, Proefshcrift, ISBN: 978-90-386-1258-4

Van der Velden, M. H. L.; Brok, W. J. M.; van der Mullen, J. J. A. M.; & Banine, V. (2006). Kinetic simulation of an extreme ultraviolet radiation driven plasma near a multilayer mirror, J. Appl. Phys. 100 (7) 073303, ISSN 1089-7550

van Herpena, M.M.J.W.; Klundera, D.J.W.; Soera, W.A.; Moorsb, R.; & Banineb, V. (2010). *Sn etching with hydrogen radicals to clean EUV optics*, Chemical Physics Letters, Vol. 484, Iss. 4-6, Pp. 197-199, (January 2010), ISSN 0009-2614

Vargas López, E.; Jurczyk, B. E.; Jaworski, M.A.; Neumann, M. J.; & Ruzic, D. N. (2005). *Origins of debris and mitigation through a secondary RF plasma system for discharge-produced EUV sources*, Microelectronic Engineering, Vol. 77, Iss. 2, pp. 95-102, (February 2005), ISSN 0167-9317

Wagner, C. & Harned, N. (2010). *EUV lithography: Lithography gets extreme*, Nature Photonics 4, 24 – 26 ISSN 1749-4885

Yoshioka, M.; Teramoto, Y.; Zink, P.; Schriever, G.; Niimi, G.; & Corthout, M. (2010). in Proceedings Vol. 7636, Extreme Ultraviolet (EUV) Lithography, 2010, 763610-1

4

Double Patterning for Memory ICs

Christoph Ludwig and Steffen Meyer
Q-CELLS SE, Bitterfeld-Wolfen
Germany

1. Introduction

In order to continue technology shrink roadmaps and to provide year by year smaller chips with more functionality, nearly all of the leading edge semiconductor companies have adopted double patterning process technologies in their fabrication lines to bridge the time until next generation EUV lithography reaches production maturity. Double patterning technology can be classified into two major main streams, however its implementation and especially the details of the process integration vary strongly among the semiconductor company and every manufacturer found his own optimum.

Therefore the given schemata will focus on the key principles. We will not stress material combinations, but point out particular challenges, show some detailed analyses, and sketch solutions.

2. Double patterning by litho-etch-litho-etch

The most straightforward approach of doing double patterning is splitting a given pattern into two parts by separating neighbouring patterns (see figure 1a). By doing this the minimum pitch of each split will be enlarged and becomes again printable with standard ARF immersion exposure tools. At the first lithography step, a photo mask containing only the blue part of the pattern will be used. As one can easily see from figure 1b, the pitch of blue only pattern is about two times larger than in the original layout. However, in order to achieve a good optical contrast during imaging as well as to get sufficient (photo resist) process window for the patterning, the pattern will usually be biased on the mask as shown in figure 1c.

  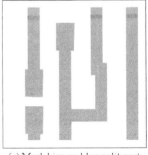

(a) Target pattern   (b) Split pattern   (c) Mask bias on blue split part

Fig. 1. Example of decomposing an irregular structure for double patterning processing.

Triangular conflicts deserve special attention. They can for instance be solved by using junctions to bridge split part features with each other as shown in figure 2. Another way is to restrict the layout to full splits and to use separate interconnects. Despite the need of such kind of additional rules, LELE can handle complex layout and modern CAD tools do support pitch splitting well.

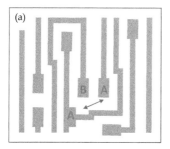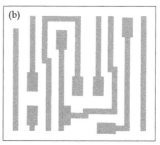

Fig. 2. Topological conflict (a) and a possible decomposition by a junction bridge (b).

To achieve the intended structure dimensions, the applied bias must be removed again. Overexposing the photo resist is commonly used to reduce the bias, but there are limits for process stability reasons. Therefore further bias reduction might be required. Often this will be done in subsequent etch steps, which also have the function of "freezing" the pattern, so that it will not be destroyed when the second part of the pattern will be added. Figure 3 shows a sample sequence.

Fig. 3. Process sequence for litho-etch-litho-etch double patterning.

The first pattern freeze is typically done in a thin sacrificial hard mask. Although this thin sacrificial hard mask adds process complexity and thus fabrication cost, there are two good arguments for introducing it. The argument #1 is as follows: It is easier to assemble all pattern elements first within a thin sacrificial layer and later transfer them altogether to the intended layer stack, rather than structuring the full intended layer stack step by step. This is because the intended structures on wafers tend to have large aspect ratios. But once the second lithography gets carried out, the topography of the wafer must be nearly flat. Otherwise resist spin-on gets problematic and one has to struggle with reflective notching and local CD (critical dimension) variation. That means, without using the sacrificial thin hard mask, complex planarization steps would be necessary. However, the thin hard mask topography differences can easily be covered by BARC / bottom layer coatings, which have to be applied for ARF resist systems anyway. The argument #2 is: Hard masks are typically needed for the plasma etch process of the intended layer stack anyway. The limits of the plasma etch selectivities and the aspect ratios require the usage of sophisticated hard mask stacks. Etching a complex stack twice is also more costly than doing the double patterning etch sequences only on the thin hard mask and perform the complex and expensive full stack etch only once. Last for the sake of completeness we want to mention here, that for cost and complexity reasons techniques of freezing the first split pattern directly in photo resist system are showing up in the industry.

Splitting the given pattern into two parts allows complex designs. Therefore this type of double patterning is the preferred choice for logic manufactures. But since two lithography steps must be done to achieve the patterning for one layer, the intra-layer overlay must be controlled with a very high degree of precision. This was challenging in the past, because first and second generation ARF immersion tools could hardly achieve overlay control beyond 6-8nm, which is above the needed budget for 36nm...32nm and 28nm...24nm technologies. Around 2010, lithography tool makers removed this blocking point by introducing a third generation of ARF immersion tools with can meet double patterning overlay control requirements. This double patterning technique is usually referenced with Litho-Etch-Litho-Etch (LELE) but other names such as Brute Force or Pitch Splitting exist as well.

For the extremely regular arrays of memory chips like e.g. NAND flash memories, aggressive double patterning technology nodes have been introduced already in 2008, before exposure tools with improved overlay control became available. Memory chip manufacturers have therefore driven another double patterning main stream, which we will describe in the next paragraph.

3. Self aligned double patterning

The second main stream is often called self aligned double patterning. Some people call this technique pitch fragmentation and you might often hear spacer based double patterning, because for most of this techniques spacer-like approaches are used to double the number of features.

Self aligned double pattering comes with strong layout limitations. However, it is well suited for regular patterns e.g. of memory chips. The design process is more complex than LELE and the pattern on the photo masks look different compared to the final pattern. This creates a first challenge, because layout versus schematic checks require additional CAD tooling to transform the self aligned double patterning mask data into the final layout data. The transformation in the other direction – from final layout to double patterning layout – is even more challenging. Automated tools are rare and manual interaction of the designers is required. This is limiting the application to a manageable number of layout features and is the main reason why self aligned double patterning is mostly only used by memory manufactures.

The big advantage of self aligned double patterning is that only one main feature lithography step has to be done. However, one or two additional assist lithography steps with considerably reduced resolution requirements are used to form the edges of the regular pattern within the memory array as well as the transition from the regular array towards the remaining area of the chip, containing the control function elements of the memory.

The lithography tool overlay requirement is only driven by the inter-layer technology demands and not by the intra-layer alignment as for LELE.

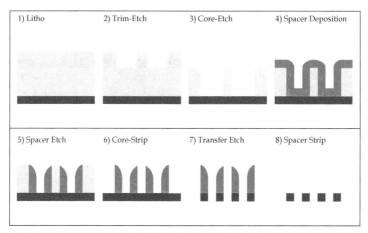

Fig. 4. Sketch of a process sequence for "line by space" double patterning.

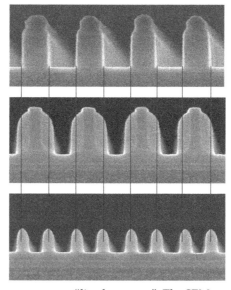

Fig. 5. Example of a process sequence "line by spacer": The SEM cross section at the top reveals the primary line pattern as defined by lithography, after print and dry etch. The cross section in the middle shows the primary lines now covered by (sacrificial) spacers, after spacer deposition and spacer etch. The bottom picture is taken after removal of the primary lines. The sacrificial spacers have become now a regular pattern of twice the spatial frequency compared to the primary pattern, and can be further used as a hard mask for the subsequent patterning of the desired stacks.

With the main lithography step, a so called "primary" or "core" pattern is printed on the wafer. This pattern has a pitch twice as large as the final structures and gets usually biased for process window reasons. Out of this pattern, a tall rectangular-shaped core structure with an aspect ratio of typical 1:3 to 1:5 is formed. This can be done directly in the photo resist system, or - more common – can be transferred by reactive ion etch into a relative thick underlying hard mask. Afterwards a thin conformal liner will be deposited onto the core shape. Now a spacer etch is performed and the core pattern gets stripped out. Finally the spacer structures will be transferred into another under laying thin hard mask and the spacer will be taken off as well. Figure 4 and 5 show such a process sequence. A detailed process description including key process parameters is listed in table 1. For this sequence the final line structure width is defined by the deposited thickness of the sacrificial spacer. Therefore this scheme is called "line by space" (LBS). All created lines will have the width of the sacrificial spacer. There is no degree of freedom for the chip layouter to modify the line width. Only the space between the lines can be varied by layout.

Process Step	Target
Deposition of CVD carbon hard mask	150nm (should to be adjusted to layer requirement)
Anneal carbon	630°C, 60min
Deposition of CVD multi-layer stack (double patterning hard mask)	25nm SiON + 25nm a-Si + 35nm SiON + 150nm a-Si + 20nm TEOS (from bottom to top)
Litho 193nm immersion	72nm line / 72 nm space CD uniformity < 3nm
Trim etch of poly mask	~ 42nm line CD uniformity < 2nm taper within the array: 90 deg
Strip Resist	
Clean Wet	minimum attack of SiON
Etch Wet TEOS	high selectivity, minimum attack of SiON
Deposition CVD HARP liner	36nm thickness thickness uniformity < 0,5nm
Etch Spacer Oxide	
Etch Recess Poly	complete carrier a-Si removal
Etch Plasma SiON	spacer transfer etch, stop on a-Si
Etch Wet Oxide	remove space; minimum attack of SiON
Subtractive Patterning	etch SiON selective to a-Si
Additive Patterning	etch aSi with SiON and resist mask with stop on SiON

Table 1. Process Sequence for Spacer Transfer Scheme (LBS) with process targets for the gate layer in a 36nm NAND flash chip.

Some applications however require lines with well-defined different line widths, but can accept equal spaces. This can be realized by a slight modification of the described processing sequence: filling the pattern after the spacer etch followed by a controlled recess etch and spacer material strip. Figure 6 demonstrates an example for a process flow which is called "line by fill" (LBF).

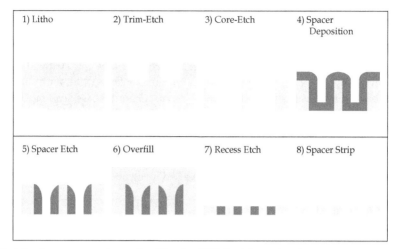

Fig. 6. Sketch of a process scheme for "line by fill" double patterning.

3.1 Line by space (LBS) – Line by fill (LBF)

The decision for either of the two basic self-aligned double patterning schemes LBS or LBF has to be taken for each lithography layer individually. In general, LBS is advantageous if the tolerances in line width are most critical for the electrical functionality of the product, and LBF is the preferred option if the distance between the lines are most critical for the product functionality. Figure 7 illustrates this difference. Therefore, sound knowledge of the fabrication process tolerances and thorough simulation are mandatory.

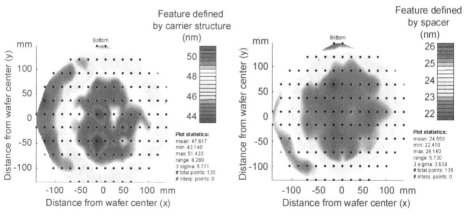

Fig. 7. Critical dimension of the gate hard mask, measured across the full wafer. The left hand graph shows the hard mask spaces, the "space-1", being defined by the primary line, whereas the right hand graph reprents the hard mask "lines", being formed by the sacrificial spacers. The hardmask lines show a much smaller spread in width than the hard mask spaces. Note that the hardmask lines are narrower than the intended gates: this is to compensate for the gate etch bias.

3.1.1 Process tolerances

One of the biggest challenges for self aligned double patterning is the control of the critical dimensions. Compared to the traditional litho, when CD control was more or less managed by lithography, self aligned double patterning relies also on tight process control for etch and deposition processes. For the example sequence sketched in figure 4, the line width is defined by the thickness of the deposited (spacer) liner and the subsequent spacer etch. Extremely low process variations, exact reproducibility, uniformity are mandatory. Therefore, special care has to be dedicated to the search of well controlled deposition processes with reasonable process cost.

The first process choice for a well-controlled deposition process could be LPCVD. However, most of the LPCVD furnace processes require process temperatures above 500°C. Front-end application might allow such temperatures, but in the back-end of line such high temperature would destroy the already manufactured metal lines or contact junctions. Therefore, LPCVD cannot be applied to critical metal layers.

PECVD processes can run at lower temperatures and come with acceptable cost. However, still most of the PECVD deposition processes are not suitable for spacer based double patterning, because the deposition mechanism is transport limited. In other words: the new material deposition reaction is very fast and is limited by how fast and how much new material can be transported out of the plasma to the wafer. The "stream" of new material is limited in time and will be distributed over the wafer surface. As a consequence the thickness of the deposited liner will vary with the surface area. Therefore dense structure areas with larger surface will see thinner liner ticknesses compared to isolated structures. Unfortunately this so-called micro-loading has quite some range and will create dozens of array edge line with different line width. As a result, PECVD processes cannot deliver tight in-chip CD control.

A highly promising process candidate is atomic layer deposition (ALD). They come with extreme good thickness uniformity and can be done at very low temperatures. The first ALD processes hat the drawback of a relatively slow process speed, resulting in high process cost. However, the original ALD with its slow atom-by-atom like deposition has experienced considerable improvements. Nowadays modified ALD processes are available like e.g. catalytic (fast) ALD, spatial ALD, molecular layer depositions or pulsed plasma depositions. Some of these advanced deposition processes can be done at temperature around or even below 100°C. This enables spacer liner depositions directly on photo resist type core materials.

The liner thickness variations are not the only critical parameter. Also for other process steps, the control of critical dimensions is challenging. The CD variation of the "primary" or "core" space results from both litho and core-etch non-uniformity. The line width variation depends on spacer deposition and etch uniformity. The "secondary" or remaining space in principle will suffer from the sum of both variations. In order to keep the "secondary" or remaining space within the allowed technology variation specifications, one has either to run each process with very tight specifications or to implement an advanced feedforward / feedback APC system. Such a system is capable e.g. of tuning the etch-bias as a function of the photo resist CD after the lithography patterning.

The uniformity of the line width is affected by the topography of the environment. There are both short-range effects and long-range effects. Transport processes and reaction kinetics of chemical and plasma reactions are sensitive to the effective surface in the vicinity of a considered line as well as to the macroscopic location on the wafer. In addition to the structure dimension of interest, monitoring structures need to be developed for a fast and reliable monitoring of the mass production.

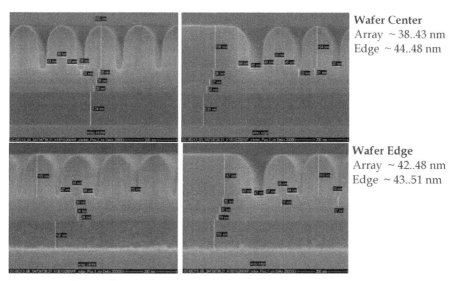

Fig. 8. Micro loading in the double patterning hard mask for the gate level of a 36nm NAND Flash. The hard mask is made for the line by space (LBS) scheme.

We have selected the gate layer of a 36nm flash memory as an example of micro loading. Figure 8 shows measurements taken from cross-sections after the formation of the sacrificial spacer at the sides of the primary lines. The hard mask is not yet structured. An in-line measurement at this process step will reveal the sum of the width of the primary line plus two adjacent spacer thicknesses, equalling roughly the desired line with plus two spaces. An intuitive way of plotting this in-line measurement number is shown in Figure 9.

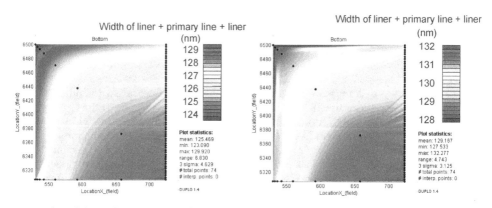

Fig. 9. Plot of the in-line measured line width consisting of the primarily formed line plus two adjacent spacers. The coloured area represents a quarter of the regular gate array. The center of the array is located at the bottom right of the plot, the corner of the array corresponds to the top left corner of the plot. Obviously, the line width increases from the center of the array towards the corners. The right hand graph was taken from an improved spacer deposition process, showing a narrower width distribution.

After removing the primary lines and transforming the sacrificial spacers into the hard mask, one obtains values as given in Table 2.

Wafer Uniformity Center to Edge	3nm
Micro Loading Array to Select Gate Edge	2..3nm
Micro Loading Gate Array Edge	6..10nm
Micro Loading Array to Litho Monitor (nested)	10nm
Micro Loading Array to Litho Monitor (isolated)	17nm

Table 2. Experimental data for uniformity and micro loading and uniformity of the the gate layer of a 36nm NAND flash patterned by LBS.

3.1.2 Gate

In general, the gate length is critical, requiring LBS (line by spacer) for the gate level. Typical electrical parameters of relevance are:
- Narrow distribution of IV characteristics of the transistors, mainly for:
- Large on-currents: short gates are preferable
- Source-Drain leakage: long gates are preferable
- Hot carrier effects: long gates are preferable

In the case of a flash memory, another electrical parameter comes into play: the coupling ratio, which determines mainly the programming and erase characteristics of the flash transistors. The coupling ratio is affected by the capacitance between the floating gate and the active area, i.e. the CD variations of both the (floating) gate and the active area cause a spread in programming and erase performance.

As a result, a carefully controlled balance of the gate length in all kinds of chips is more important than the control of the gate spaces.

3.1.3 Active area / STI

For the active area, the following elements are relevant:
- Transistor on-current: active area width variations translate directly into transistor current variations.
- Coupling ratio of flash transistors: active area width variations translate into the program and erase speed.
- Transistor leakage: The STI (shallow trench isolation) fill turns out to be highly sensitive to the STI trench width. Trap centers at the boundary of transistor channel edge and STI fill can result in undesired leakage currents, both along the channel direction and across the source/drain junction into the substrate.

The balance of the mentioned three arguments led the authors to the decision of LBF (line by fill) for the active area layer.

3.1.4 Bit line contact

NAND-Flash memory arrays require single rows of so called dense bit line contacts. They connect the metal bit lines to the active area, and they are following the critical pitch. The special challenge consists in the lack of any room for overlay tolerances and in the extreme asymmetry of the pattern: the contact row can be considered an isolated structure in one dimension, and in the orthogonal dimension a periodic structure with minimum pitch.

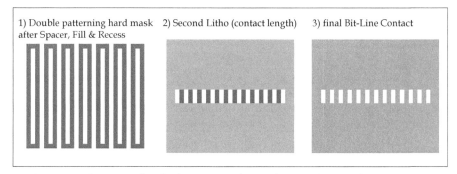

Fig. 10. A two-mask process for the formation of a single row of on-pitch contacts.

Figure 10 shows one example how to realize elongated dense bit line contacts. In a first step a dense array-like hard mask is created by double patterning. The lines and spaces of this array-like hard mask define the bit line contact widths and the contact to contact distances, respectively. The array is much larger than the intended contacts in order to reach a good imaging in the lithography tool. A second lithography layer is used to cut out only a small region of the array-like hard mask. This second litho step prints a long opening in a resist mask across the array-like hard mask. The width of this long opening defines the length of the contacts. The following etch process is to open only the regions which are neither covered by the spacer nor by the photo resist.

3.1.5 Tight pitch metal layer
The predominant issues for the selection of the proper double patterning scheme for metallization are
- dielectric breakthrough,
- variations of capacitive coupling,
- variations of resistance,
- feasibility of landing pads for contacts / vias for the electrical connection to layers beyond or beneath.

Typically, bit lines are supposed to run fast electrical signal pulses with lowest possible loss and lowest possible cross-talk. This requirement suggests that a lower limit of the metal space might be important in order to minimize RC losses. Simulations have been performed for the bit line capacitance in a flash memory for both the LBF (line by fill) and the LBS (line by space) process scheme. The results led to the conclusion that both LBS and LBF show nearly identical variations with respect to the capacitive coupling between adjacent bit lines. The resistivity variations, however, are clearly in favour of the LBS scheme, which has a well controlled metal line width, i.e. the resistance tolerances are much better than for LBF. But in the end, deeper study of process architecture and complexity and cost led to the decision of LBF.

Two-dimensional electric field and capacitance simulations have been performed for LBF (figure 11) and LBS. The calculations include fringing fields for up the third neighbour bit line as well as the capacitance to the top and bottom layers of the chip. Worst and best cases, i.e. largest and lowest capacities have been calculated. For the variations, not only the double patterning induced line and space variations have been considered, but also variations in bit line thickness. The target thickness of the considered example is 60nm for the full metal stack, the tolerance is +/- 10nm.

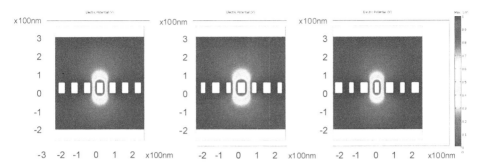

Fig. 11. Electric field simulations for bit line capacitance calculations, depicted for the LBF process.

It is not upfront clear by intuition, which combination of double patterning variations will result in the largest and lowest bit line capacitance, respectively. For example, in the LBS scheme, the smallest bit line capacitance does not occur for the largest primary space. The reason is the linear (inter-)dependence of the four parameters: The sum of two spaces and two lines is fixed to twice the pitch, 144 nm in our example. An illustration of the line / space variations towards lower capacitances is given in figure 12. The lowest possible value for LBS (3 sigma geometrical variations assumed) is 2,01 x 10^{-10} F/m, which is 10% below the value of the target geometry (2,23 x 10^{-10} F/m).

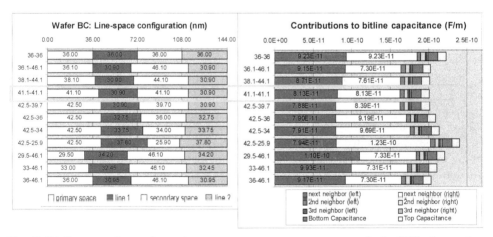

Fig. 12. Bit line capacitance for various extreme geometry parameters in the LBS scheme, for lowest capacitances from process variations ("best case"). A totally symmetric configuration of 41.1 nm for both the primary and the secondary space and a line width of 30.9 nm is found to be the best case, resulting in a total capacitance of 2,01 x 10^{-10} F/m.

A summary of the results is depicted in figure 13 and listed in table 3. The outcome of the considerations is a tiny preference for the LBS scheme.

The feasibility of process architecture finally requires a clear decision between the two double patterning schemes. A reliable connection to the bit line contact and a robust process margin require some kind of so called landing pad, i.e. a local enlargement (landing pad) of the bit line around the bit line contact. This is only feasibly by LBF, since only in the LBF, the layouter can intentionally modify the width of the bit line. The LBS would ascribe the bit lines a well-defined, fixed width.

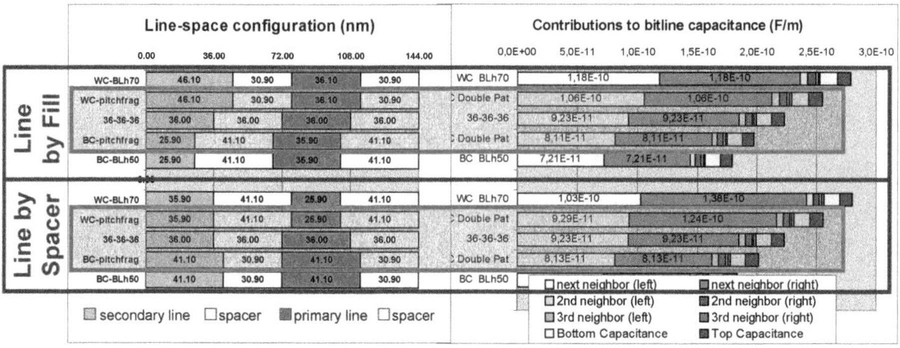

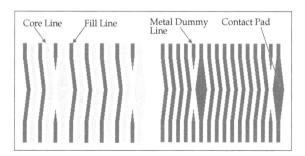

Fig. 13. Bit line capacitance in a 36nm flash M0 layer for the extreme 3 sigma cases of double patterning variations, marked by the red boxes, and in addition bit line height variations of +/- 10nm.

	LBF (10⁻¹⁰ F/m)	LBF (%)	LBS (10⁻¹⁰ F/m)	LBS (%)
WC incl. BL height (3 σ)	2,61	17	2,61	17
WC (double patt. only) (3 σ)	2,54	14	2,55	14
Target dimension	2,23	0	2,23	0
BC (double patt. only) (3 σ)	1,97	-12	2,01	-10
BC incl. BL height (3 σ)	1,90	-15	1,93	-14

Table 3. Bit line capacitance (numbers absolute and relative to the geometrical target dimensions) for LBF and LBS. All variations represent 3 sigma values. The range between the largest, i.e. worst case (WC) capacitance and the lowest, i.e. best case (BC) capacitance is roughly identical for both double patterning schemes. Besides the pure double patterning effects, the table also considers a calculation of the bit line height variations. Altogether, the two double patterning schemes do not exhibit a significant difference.

Core Line   Fill Line   Metal Dummy   Contact Pad
                                Line

Fig. 14. LBF allows local enlargements of the bit line width.

In an NAND flash, the regular bit line pattern exhibits periodic interruptions which are needed for the source line contacts. This enables the introduction of some dummy extra bit lines. They are used to allow tiny landing pads on top of the bit line contacts. Figure 14 shows an example for bit line landing pads.

3.2 Fan-out

So far, we have focussed our discussion mainly on the regular array. The edge of the array deserves special attention.

Self-aligned double patterning naturally creates lines or spaces which are connected pair-wise. In general, this is not desired. Therefore after the array formation by double patterning, the edge of the array is cut by applying a second uncritical litho & etch step, as illustrated in figure 15. Preferably, this step is applied to the hard mask before etching the full stack.

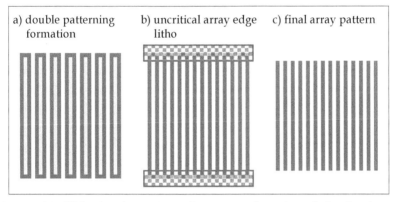

a) double patterning formation   b) uncritical array edge litho   c) final array pattern

Fig. 15. An uncritical litho & etch step is used to remove the unintended pair-wise connections of the double patterning lines.

The array lines have to be connected electrically to other functional elements on the chip. This requires a so-called fan-out, a topological connection between the double patterning elements and the structures printed by direct lithography.

A possible process sequence of a "Christmas tree" fan-out for the gate layer of a NAND flash is shown in figure 16. The ends of the lines formed by primary lithography are drawn in a shape as depicted in figure 16 (a). The first lithography step prints the primary lines, having twice the array pitch, and forms the fan-out core. The spacer covering the sidewalls of the primary lines follows the core edges and routes each array line into the Christmas tree fan out (figure 16 (b)). The spacer has the intended pitch within the array and is transformed into array lines. They are connected in pairs. Therefore an additional (uncritical) litho & etch cut step must be performed (figure 16 (c)). Finally, landing pads are added for better overall process robustness (figure 16 (d)). The additive structuring requires one more litho & etch, which is usually combined with the periphery patterning. For the additive structuring, the combination of the already formed double patterning hard mask and the add-on photo resist act as an etch mask for the final etch.

A SEM photography of such a "Christmas tree" word line fan-out is given in figure 17.

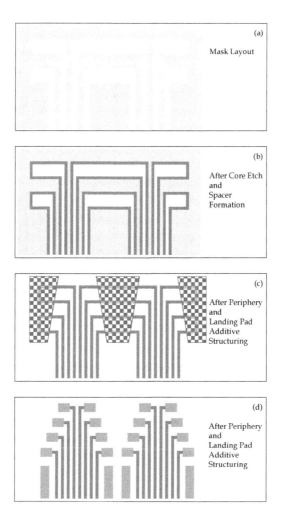

Fig. 16. "Christmas tree" fan-out for the gate layer of a NAND flash.

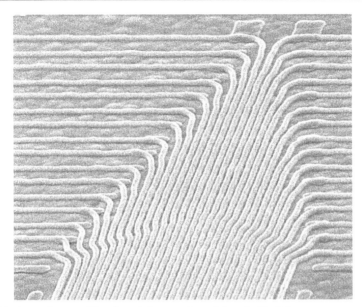

Fig. 17. SEM view of a word line (gate) fan-out for a 36nm half pitch array.

4. Conclusion

Double Patterning has become an important technique for advanced microelectronics devices. Cost-competitive memory chips have structure dimensions which are well below the diffraction limit of any available productive lithography tool. The generation EUV lithography is still on its way towards production maturity.

We have shown a classification of double patterning and have discussed possible implementations. Their specific challenges and advantages have been considered. For a 36nm NAND flash as an example, process flows have been presented.

5. Acknowledgment

This work was financially supported in part by the European Commission in the projects GOSSAMER and PULLNANO, and by the Federal Ministry of Education and Research of the Federal Republic of Germany (project nos. 01M3167 A and 01M3171 A).
The authors would like to express their gratitude to the whole flash development team of the former company Qimonda, especially Tim Höhr for the capacitance simulations.

6. References

Beug, M.F.; Hoehr, T.; Müller, T.; Reichelt, R.; Müller-Meskamp, L.; Geiser, P.; Geppert, T.; Bach, L.; Bewersdorff-Sarlette, U.; Kenny, O.; Olligs, D.; Brandl, S.; Marschner, T. ; Parascandola, S.; Meyer, S.; Riedel, S .; Specht, M.; Manger, D.; Knöfler, R.; Knobloch, K.; Kratzert, P.; Ludwig, C. & Küsters, K.-H. (2008). Pitch Fragmentation Induced Odd/Even Effects in a 36nm Floating Gate NAND Technology,

International Electron Devices Meeting Technical Digest, p.353, San Francisco, CA, USA, December 15-17, 2008, ISSN 8164-2284

Beug, M.F.; Knöfler, R.; Ludwig, C.; Hagenbeck, R.; Müller, T.; Riedel, S.; Höhr, T.; Sachse, J.-U.; Nagel, N.; Mikolajick, T.; Küsters, K.-H. (2008). Charge cross talk in sublithographically shrinked 32nm Twin Flash memory cells, *Solid-State Electronics Vol.* 52, pp. 571-576

Ghaida, R.S.; Torres, G.; Gupta, P. (2011). Single-Mask Double-Patterning Lithography for Reduced Cost and Improved Overlay Control, *Semiconductor Manufacturing,* Volume: 24 , Issue:1

Dusa, M.; Arnold, B.; Finders, J.; Meiling, H.; van Ingen Schenau, K.; Chen, A. C. (2008). The lithography technology for the 32nm HP and beyond, *Proceedings of the SPIE,* Volume 7028, pp. 702810-702810-11, DOI: 10.1117/12.796016

Bencher, C.; Chen, Y.; Dai, H.; Montgomery, W.; Huli, L. (2008). 22nm half-pitch patterning by CVD spacer self alignment double patterning, *Proceedings of the SPIE,* Volume 6924, pp. 69244E-69244E-7, DOI: 10.1117/12.772953

Wie,Y.; Brainard, R. L.. (2009). Advanced Processes for 193-nm Immersion Lithography Book, ISBN: 0819475572

Chiou, T.-B.; Socha, R.; Kang, H.-Y.; Chen, A. C; Hsu, S.; Chen, H.; Chen, L.. (2008). Full-chip pitch/pattern splitting for lithography and spacer double patterning technologies, *Proceedings of the SPIE*, Volume 7140, pp. 71401Z, DOI: 10.1117/12.804763

Noelscher, C.; Jauzion-Graverolle, F.; Heller, M.; Markert, M.; Hong, B.-K.; Egger, U.; Temmler, D. (2008). Double patterning down to k1=0.15 with bilayer resist, *Proceedings of the SPIE*, Proceedings of the SPIE, Volume 6924, pp. 69240Q-69240Q-12, DOI: 10.1117/12.772750

Ludwig, C ; Beug, M.F. ; Küsters, K.-H. (2010). Advances in Flash Memories, *Materials Science-Poland*, Vol. 28, No. 1. 105-116

5

Diffraction Based Overlay Metrology for Double Patterning Technologies

Prasad Dasari[1], Jie Li[1], Jiangtao Hu[1],
Nigel Smith[1] and Oleg Kritsun[2]
[1]Nanometrics
[2]Globalfoundries
USA

1. Introduction

193nm optical immersion lithography is approaching its minimum practical single-exposure limit of 80nm pitch [1]. The semiconductor industry has adopted double patterning technology (DPT) as an attractive solution for the low k_1 regime until extreme ultraviolet (EUV) lithography becomes commercialized. DPT also brings additional demands of increased critical dimension uniformity (CDU) and decreased overlay errors. The International Technology Roadmap for Semiconductors (ITRS) [2] target for overlay control at the 32nm DRAM node in single patterned lithography steps is 6nm. The process budget is reduced to 1.1nm for DPT. If 20% of the process error budget is allowed to occur in the metrology tool, as the ITRS states, then the measurement error budget at the 32nm node is 1.2nm for single patterning, and 0.22nm for DPT.

The ITRS defines total measurement uncertainty (TMU) for overlay only in terms of precision, tool-induced shift (TIS) variation and site-to-site tool matching differences. Determining whether a measurement technology is capable of controlling these advanced processes is no longer a case of simple data self-consistency checks on precision, TIS and matching. For example, the error arising from assumptions of a linear change of overlay error with position is significant. This error can be reduced by using very small targets [3] and performing in-device overlay measurements, but the demanding sub-nanometer measurement budget in overlay measurements still remains a considerable challenge.

Recent advances in lithography metrology for advanced patterning have led to the proposal of three different pitch splitting technologies [Fig. 1]. The Litho-Etch-Litho-Etch method (LELE, Fig. 1a) involving two process steps requires very tight overlay control and is both very expensive and slow, making alternative methods attractive. The first alternative process flow is Litho-Freeze-Litho-Etch (LFLE), which reduces the processing cost by replacing the intermediate etch step with a process step in the litho track (Fig. 1b). After exposing the first pattern, the resist is baked in a post-exposure bake (PEB) step and developed. Exposed pattern is coated with material to freeze the resist. The second resist layer is added and the second exposure is done. The freezing material prevents the first resist layer from washing away during the second layer PEB and develop steps. This technique allowed printing 2D logic cells and dense poly lines with two lithography steps, illustrating good resolution and process margin [4].

The next alternative process is Self-Aligned Double Patterning (SADP, Fig. 1c), in which a spacer film is formed on the sidewalls of pre-patterned features. Etching removes all the material of the original pattern, leaving only the spacer material. Since there are two spacers for every line, the line density has now doubled. The spacer approach is unique in that with one lithographic exposure the pitch can be halved indefinitely with a succession of spacer formation and pattern transfer processes. The spacer film deposition process is very uniform and results in extremely good SADP CDU of less than 1nm. The spacer lithography technique has most frequently been applied in patterning fins for FinFETs and metal layers [5].

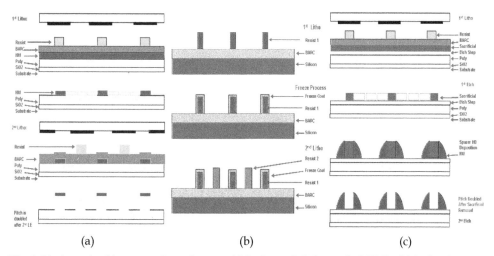

(a)   (b)   (c)

Fig. 1. Various double patterning schemes: (a) Litho-etch-litho-etch (LELE), (b) Litho-freeze-litho-etch (LFLE), and (c) Self-aligned double patterning (SADP)

These pitch splitting double patterning techniques not only involve more demanding process steps, they also require tighter overlay control than conventional single patterning [2]. Therefore measurement of overlay with much higher certainty is a necessity. As technology transitions toward the 22nm and 16nm nodes using these methods there is serious concern about the capability of the available metrology solutions, both in process development and production control.

High TIS and tool-to-tool matching errors make it difficult to meet the measurement uncertainty requirements using the traditional Image-Based Overlay method (IBO), even though most advanced IBO tools are operating at TMU levels under 1nm. Diffraction-based (scatterometry) overlay (DBO) measurement is an alternative optical measurement technique that has been reported to offer better precision than IBO and near zero TIS [6, 12, 14-15], and is therefore a possible solution to the measurement uncertainty budget. Bischoff *et al.* proposed measuring overlay using the diffraction efficiencies of the first diffracted orders [7]. Chun-Hung Ko used angular scatterometry combined with an experimental library to determine the overlay error on ADI stacks with intermediate poly-silicon lines [8]. H.-T. Huang *et al.* used spectra from reflection symmetry gratings and a rigorous coupled-wave analysis (RCWA) regression approach to calculate the overlay error [9]. W. Yang *et al.* [10] and D. Kandel *et al.* [11] used arrays of specially constructed pads with programmed offsets to determine overlay without the need for model fitting. These DBO methods have the potential to meet the demanding overlay metrology budget for sub-32nm technology nodes. In this chapter, the advantages of DBO for precise and accurate overlay measurement in LELE, LFLE and SADP processes will be shown.

2. Spectroscopic scatterometry

2.1 Experimental setup

Spectroscopic scatterometry is used to measure overlay errors between stacked periodic structures (*e.g.*, gratings). In this technique, broadband linearly polarized light is incident perpendicular to the wafer surface and the zero-order diffracted signal (spectrum) is measured as a function of wavelength. Fig. 2 shows a typical experimental configuration. At normal incidence, different reflectance spectra are obtained for various angles of polarization with respect to that of the periodic structure. Typical data collection involved both TE and TM spectra. A specific advantage of using polarized light is that it provides enhanced sensitivity as both the amplitude and phase differences between the TE and TM spectra can be measured.

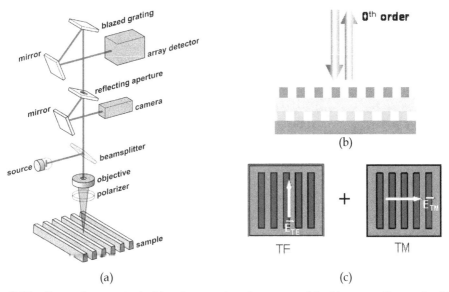

Fig. 2. The figure shows a typical hardware set up for a normal incidence scatterometer (a) spectroscopic reflectometry, (b) normal incidence reflection, and (c) TE TM data acquisition modes.

2.2 Theory

Spectra are obtained from pads, each of which has gratings patterned in both layers between which the overlay error is being measured (Fig. 3). The gratings in each pad are overlaid but by design shifted with respect to each other. Spectra from pads with shifts of equal magnitude but opposite direction are identical due to symmetry:

$$R(+x_0, \lambda) = R(-x_0, \lambda) \tag{1}$$

Here $R(x_0, \lambda)$ is the reflectance spectrum from one pad as a function of wavelength λ and shift $+x_0$. The difference spectra $(\Delta R(\lambda) = R(+x_0, \lambda) - R(-x_0, \lambda))$ from two pads with shifts $+x_0$ and $-x_0$ is zero in the absence of noise in the measuring tool. A small overlay error shifts both upper gratings in the same direction and breaks the symmetry. The resulting differential spectrum is proportional to the direction and magnitude of the overlay error:

$$\Delta R(\lambda) = R(x_0 + \varepsilon, \lambda) - R(-x_0 + \varepsilon, \lambda) = R(x_0 + \varepsilon, \lambda) - R(x_0 - \varepsilon, \lambda) \cong 2\varepsilon \frac{\partial R}{\partial x}\bigg|_{x_0} \tag{2}$$

Here, ε is the overlay error, x_0 is the offset bias, and a Taylor expansion of the reflectance around x_0 has been applied. The overlay error can now be calculated by comparing the measured differential spectrum, ΔR, to a second differential spectrum, $\Delta R'$, acquired from a pair of test pads having a known relative offset. If, for example, a shift of $x_0 + \delta$ is designed into a third pad, then within the linear-response range the difference between its spectrum and that from the $+x_0$ pad is:

$$\Delta R'(\lambda) = R(x_0 + \delta + \varepsilon, \lambda) - R(x_0 + \varepsilon, \lambda) \cong \delta \frac{\partial R}{\partial x}\bigg|_{x_0} \tag{3}$$

Equation (3) provides the calibration required to calculate the overlay error, ε:

$$\varepsilon = \frac{\delta \Delta R(\lambda)}{2\Delta R'(\lambda)} \tag{4}$$

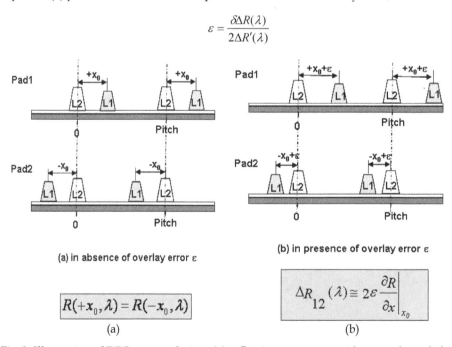

(a) in absence of overlay error ε

(b) in presence of overlay error ε

$$\boxed{R(+x_0, \lambda) = R(-x_0, \lambda)}$$

(a)

$$\boxed{\Delta R_{12}(\lambda) \cong 2\varepsilon \frac{\partial R}{\partial x}\bigg|_{x_0}}$$

(b)

Fig. 3. Illustration of DBO targets design: (a) reflection symmetry with $+x_0$ and $-x_0$ shift (no spectral difference), (b) reflection symmetry broken due to overlay error with $+ x_0 + \varepsilon$ and $- x_0 + \varepsilon$ shift (ΔR_{12} spectral difference between pads 1 and 2).

From equation (4), the ratio $\Delta R(\lambda)/\Delta R'(\lambda)$ must be independent of λ. This arises from our assumption that ε is small and only applies in the linear-response range where this assumption holds.

Equation (4) shows that the overlay can be measured in one direction using a minimum of three pads with suitably defined offsets. In practice four pads are often used (Fig. 4a), with an additional calibration pad with offset $-x_0-\delta$, because the additional data improves precision and provides a check that the overlay error is within the linear response range (Fig. 4b). To measure overlay in two directions two sets of gratings are required. The second set with the gratings rotated by 90° from the first. As overlay is a vector quantity, it is usually measured in both X&Y directions. The nomenclature "2x3 target" and "2x4 target" indicates whether three or four pads are used to measure in each axis.

The nomenclature CD/pitch is also used to indicate the designed CD and pitch of the gratings in each target. For example 65/390 means CD=65nm and pitch=390nm. All targets use the same CD and pitch at both layers.

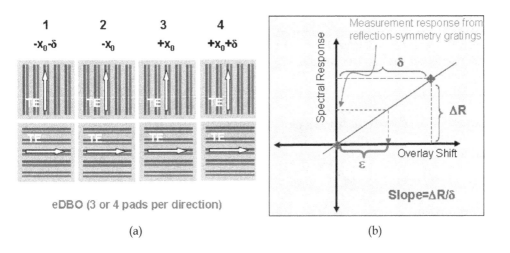

(a)   (b)

Fig. 4. (a) 2x4 target used for measurement of signal by a normal incidence scatterometer and (b) linear dependency of overlay shift as a function of spectral response.

2.3 Spectral response to overlay

Fig. 5(a) and (b) show the spectral response to overlay in the difference spectra from programmed reflection-symmetry gratings (Equation 2). The black line in the figure is the average $\Delta R'(\lambda)$ from both pairs of calibration pads, scaled by $(2\varepsilon/\delta)$. The scaled signal shows excellent agreement with the measured response, in accordance with equation 4. The maximum response increases from ~15x (Noise) to ~45x (Noise) for a ~3x change in overlay error. As expected, the spectral response also changes in sign with the measurement. The maximum spectral response at any wavelength is about 4x (Noise) per 1 nm overlay error. Measurement uncertainty much better than 0.25nm is possible because the data is summed over all available wavelengths.

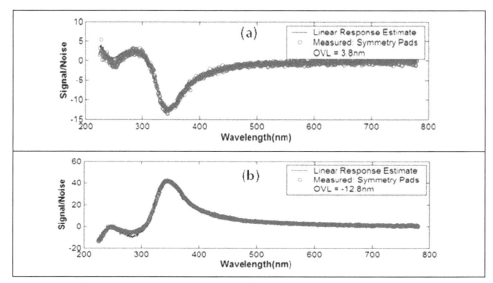

Fig. 5. Spectral response (signal to noise ratio) for corresponding overlay errors (shown in the legend box). The plots (a) and (b) also show response (black line) calculated from the calibration pad spectra.

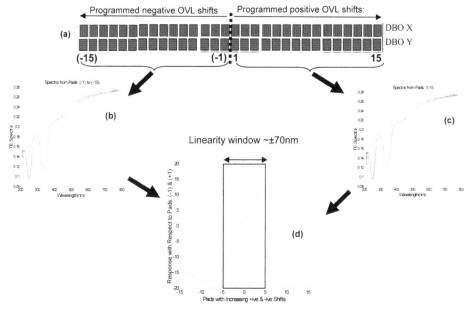

Fig. 6. (a) shows Pads 1-15 with increasing programmed positive overlay shifts and Pads (-1)-(-15) with increasing negative overlay shifts, (b) and (c) show TE spectra from corresponding vertical and horizontal gratings pad sequence respectively, and (d) shows measured spectral response in arbitrary units.

2.4 Range of linearity

The linearity range is tested by printing a sequence of pads with varying overlay shifts. Pads in the right half of the sequence (pads 1-15) have increasing positive overlay shifts (Fig. 6(a)) in 15 nm steps. Pads (-1) to (-15) have increasing negative overlay shifts in 15 nm steps. Fig. 6b and 6c shows raw TE spectra collected from these pads. The difference signal is calculated by subtracting the pad 1 spectrum for pads 2-15, and the pad -1 spectrum for pads -2 to -15. The range of linearity observed (~±70 nm) is significant for this process as shown in Fig. 6(d). A similar linearity range is observed for the horizontal gratings.

3. Litho-etch-litho-etch (LELE)

In the absence of a EUV processing solution below 80nm pitch, DPT using the litho-etch-litho-etch dual step became an attractive solution for the low k_1 regime.

3.1 DPT structures for testing DBO

The first DPT test structure is on a silicon substrate (fig 7a). The structure consists of ~120 nm photoresist lines and ~40 nm nitride lines with silicon over etch. The second DPT stack (Fig. 7b) comes from a gate (bitline)-level patterning step in a NOR flash process.

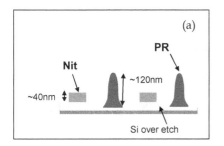 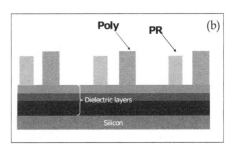

Fig. 7. (a) DPT test structure with ~120 nm photoresist lines and ~40 nm nitride lines with silicon over etch on a silicon substrate. (b) Gate level (bitline) patterning step in a NOR flash process.

3.2 DBO Measurement accuracy: correlation with IBO and CD-SEM

DBO measurement accuracy is assessed by comparing the results against IBO and CD-SEM data.

3.2.1 Correlation with IBO data

Fig. 8(a) shows the comparison between DBO and IBO measurements on all 143 fields on the wafer. There is a very good correlation between DBO and IBO measurements (R^2=0.99) with an offset of ~7 nm. The correlation is good for the subset of measurements less than ±3 nm (Fig. 8b). The inset histogram in Fig. 8a shows the difference in IBO and DBO measurements after removing the ~7 nm constant offset. The distribution is approximately normal with standard deviation of 1.8nm.

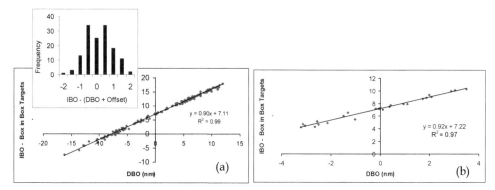

Fig. 8. DBO vs. IBO correlations: (a) shows correlations on 143 targets throughout the wafer, and (b) shows a subset of the same data in a narrow range of ~±3 nm.

3.2.2 Correlation with CD-SEM

Fig. 9a shows a CD-SEM image of the nitride (dark) and photo resist (gray) lines. Fig. 9(b) and 9(c) show the correlation between DBO and CD-SEM measurements. In Fig. 9b, the overlay errors are calculated from the CD-SEM data for the top of the lines, while in Fig. 9c, the bottom CD data is used. The inset histograms in Fig. 9b and 9c show the distributions of the difference between IBO and CD-SEM measurements after subtracting offsets.

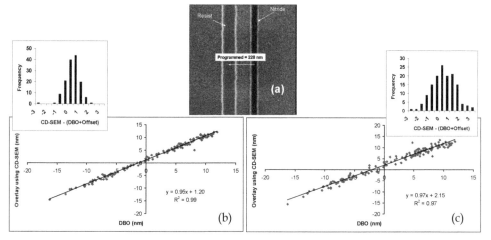

Fig. 9. (a) CD-SEM image of the DPT structure. Dark lines are nitride and the light gray lines are the photo resist, (b) correlations between DBO and top CD-SEM measurements from all 143 fields. The inset histogram shows the histogram of differences between the CD-SEM and DBO results. (c) as (b) but using the bottom CD data.

The slope of the CD-SEM/DBO correlation is 0.95 (for top CD) and 0.97 (bottom CD) compared to 0.9 for the DBO-IBO correlation. The offset between DBO and CD-SEM measurements is in the order of 1 to 2 nm. Concerns of accuracy of IBO measurements have been raised before. Dusa *et al.* reported an offset of ~5 nm between CD-SEM and IBO for a

DPT application [10]. It is possible that this offset might be coming from differences in mask writing errors between the DBO and IBO targets. The CD-SEM results are closer to the DBO values (Fig. 9), which suggests that the accuracy of the DBO technique is better than that of IBO. The observed DBO repeatability is better than CD-SEM. DBO is also non-destructive, as the sample is not subjected to the charging effects that occur in a CD-SEM.

3.3 DBO target types – 4 vs. 3 pads

While scatterometry offers precise and accurate overlay measurements for DPT, the number of reference and sample pads required for such measurements are still a concern. In this section the possibility of reducing the number of pads required without sacrificing the performance is explored using two target types – 2x4 and 2x3 targets (see definitions of target types in Section 2). Fig. 10 shows excellent correlation between overlay measured from these 2 target types. Table 1 shows the root mean square dynamic precision on two 2x3 targets and a 2x4 target. The precision on the 2x3 targets is a factor of ~1.4 (= $\sqrt{2}$) higher than that on the 2x4 target because the number of difference spectra used in the 2x4-target algorithm is twice the number used for the 2x3-target algorithm. The figure shows excellent correlation, suggesting that a 3-pad target is sufficiently accurate for this application. The dynamic (load-unload) and static (no unload) precision is excellent.

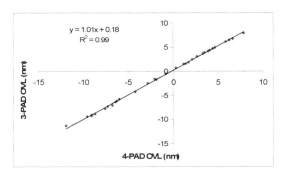

Fig. 10. Correlation between DBO targets with 4 pads/direction and 3 pads/direction.

Target/Measurement Type	RMS Precision 3σ (nm)
2x3 Dynamic, Target-II	0.31
2x3 Dynamic, Target-II	0.32
2x4 Dynamic, Target II	0.20
2x4 Static, Target II	0.11

Table 1. Shows the dynamic precision for the 2x3 target is higher by a factor of ~ $\sqrt{2}$

3.4 DBO performance: precision, TIS, matching, and TMU

DBO capability was further assessed by calculating the measurement uncertainty (TMU) from the measurement data using equation (5). Short-term (dynamic) precision ($DYNP$) is the Root-Mean-Square (RMS) of three times the standard deviation of the 20-cycle measurements at each target, and combines static precision (occurring without any movement within the tool) and the effects of target reacquisition. $TIS3\sigma_T$ is three times the standard deviation of the TIS in the measurements at each target. Where multiple tools are available, site-by-site tool matching (TM) is included. TM is three times the standard deviation of the difference in the average measurement from each tool at the same location. Matching data is not included in the DBO to IBO TMU comparison.

$$TMU = \sqrt{(DYNP)^2 + (TIS3\sigma_T)^2 + (TM)^2} \qquad (5)$$

The results for both DPT applications are summarized in Table 2. For the gate level DPT the dynamic precision is less than 0.1 nm and the TMU is 0.26 nm Average TIS is under 0.1nm. DBO matching data is between two tools (Atlas and FLX) with similar reflectometer optical heads. For the DPT structure on a silicon surface precision is 0.2nm. TIS and matching data is not available for this structure.

TECHNOLOGY	Process Step	DYNP (nm)	TIS Avg (nm)	TIS 3σ (nm)	Tool Match	TMU
DBO	DPT Silicon Substrate	0.2				
	DPT Gate Patterning NOR Flash	0.07	-0.04	0.17	0.18	0.18
IBO	DPT Silicon Substrate	0.48	-0.37	0.31		0.57
	DPT Gate Patterning NOR Flash	0.33	-2.05	6.03		6.04

Table 2. Performance summary of DBO on two stacks discussed in Fig. 4.

4. Litho-freeze-litho-etch (LFLE)

While LELE involving two process steps offers an adequate solution for DPT process steps, both are very expensive and slow. The alternative Litho-Freeze-Litho-Etch (LFLE, Fig. 1b) process reduces cost by replacing the intermediate etch step with a process step in the litho track.

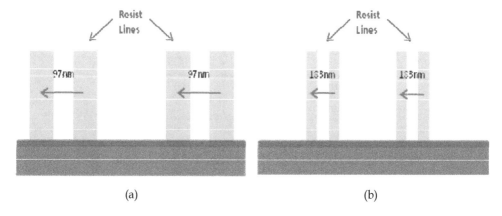

(a)   (b)

Fig. 11. DPT structure that has alternative photo resist lines with silicon over etch: (a) 65/390 Line/Pitch ratio, (b) 110/660 Line/Pitch ratio

4.1 DPT Structure
An example LFLE DPT structure consists of ~120 nm photoresist lines with silicon over etch as shown in Fig. 11(a) and (b).

4.2 Prediction based on simulation
The LFLE stack is built on silicon with BARC and two resist lines patterned on top. Simulated spectral response curves (for example from NanoDiffract™, Nanometrics scatterometry software) are used to predict the static precision of overlay measurements in the range of interest for the LFLE DPT stack.

Static precision for the LFLE model can be determined by two different methods. The first method uses analysis of measurement covariance as described by Vagos *et al* [16], and which is referred to as the "Uncertainty and Sensitivity Analysis" method (U&SA). In the second method random noise is introduced into the simulated DBO spectra for all four pads and static precision determined as if the spectra are obtained experimentally. The predicted static precision for the LFLE stack is 0.24nm (3σ) using the U&SA method and 0.30nm by the noise induced method.

4.3 DBO accuracy (freeze process)
4.3.1 Correlation with IBO
The CD-SEM image in Fig. 12a shows the resist lines in one of the four pads of a 110/660 target. To test measurement accuracy, the results from two 2x4 DBO targets (65/390, 110/660) are compared with IBO measurements using nearby Blossom targets. Fig. 12b shows the excellent correlation (R^2 ~0.99) between the DBO and Blossom data.

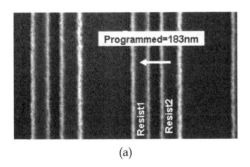
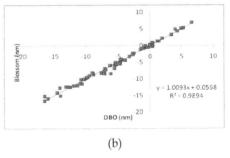

(a)   (b)

Fig. 12. Correlation of DBO vs. IBO: (a) 110/660 pad, (b) Correlation of DBO 2x4 targets vs. blossom

4.4 LFLE DBO performance
Good overlay control requires good measurement capability. Table 3 summarizes the precision, TIS, tool matching and measurement uncertainty (TMU) for DBO of LFLE structures. DBO dynamic precision was less than 0.2nm and TMU less than 0.5nm. Average TIS was under 0.1nm.

Table 3 contains site-by-site (SBS) 3σ DBO matching data for this stack from three tools with the same design of reflectometer optical head. Tool matching of 0.14nm or less is achieved without calibration or adjustment of the tools. This is possible because the method of equation (4) is self-calibrating. Absolute spectral matching between tools is not necessary.

TECHNOLOGY	CD:Pitch	Tool	DYNP (nm)		TIS Avg (nm)		TIS 3σ (nm)		Tool Match		TMU	
			X	Y	X	Y	X	Y	X	Y	X	Y
DBO	65:390	Tool1	0.19	0.17	0.06	-0.01	0.42	0.44	0.09	0.14	0.47	0.49
	65:390	Tool2	0.18	0.18	-0.07	0	0.36	0.37	0.09	0.14	0.41	0.44
	110:660	Tool2	0.21	0.18	0.07	0.02	0.28	0.32	0.06	0.07	0.36	0.38
	110:660	Tool3	0.21	0.19	-0.08	-0.08	0.21	0.19	0.06	0.07	0.3	0.28
IBO		Tool4	0.71	0.59	0.18	-0.13	0.56	0.31			0.9	0.67

Table 3. Performance summary of DBO on LFLE stack.

4.5 Prediction vs. observation
The dynamic precision (3σ) of ~0.20nm is slightly better than the predictions made in section 4.2 of 0.24nm (U&SA method) and 0.30nm (noise induced method).

5. Spacer double patterning

While LFLE minimizes the number of process steps needed and thus reduces cost it still requires very tight overlay control. Spacer (SADP) forms lines around pre-patterned features, relaxing the requirement for overlay control and potentially allowing the indefinite pitch halving.

5.1 Spacer 1st layer patterning
In this section we discuss some results from first layer patterning by SADP. Experimental TE and TM spectral data obtained for an SADP stack (Fig. 13a) is fitted to modeled spectra (Fig. 13b). Although the model has not been optimized to improve the fit there is good agreement between the modeled and actual spectrum.

The CD-SEM image in Fig. 13c shows the spacer pattern. The measured bottom CD is around 48nm for layer 1 spacers. Although the spacer structures are expected to be identical for both layer 1 and layer2, the CD may vary depending on the application. The model fitting data (Fig. 13a) is consistent with uniform spacer formation across the wafer. The spacer width at the bottom is 42.5 ±1nm and SWA is 72.3 ±0.2 deg.

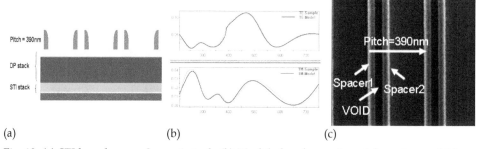

(a)   (b)   (c)

Fig. 13. (a) STI based spacer Layer 1 stack, (b) Modeled and experimental spectra, and (c) CD-SEM Image of spacer L1.

5.2 Prediction of spacer 2ⁿᵈ layer precision

The second spacer layer patterning characteristics were predicted using simulation. Since the static precision predicted for LFLE (in sections 4.2 and 4.4) was consistent with the experimental data in table 3, predictions for the double layer SADP case should also be valid. The good fit between modeled and experimental TETM spectra in Fig. 13b further supports the validity of this approach.

The final SADP stack after completion of the patterning steps is shown in Fig. 14. For LELE and LFLE, both patterning steps are done on the same layer. In the spacer case, the first patterning step is done on L1 (STI+DPT) and second patterning step is done on L2 (WL + DPT) with a programmed shift. The CD of the two spacers can be different.

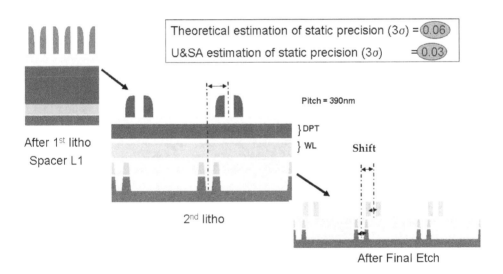

Fig. 14. The spacer stack DPT steps shown for pitch of 390nm and bottom CD ~65nm

The simulated TETM spectra for the spacer stack in Fig. 15(a) shows sensitivity across the spectral region. The TETM spectral response to overlay shift in Fig. 15(b) is linear for overlay around 25% of the grating pitch.

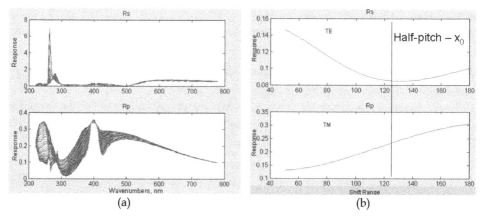

(a)   (b)

Fig. 15. (a) RCWA simulated spectra; (b) DBO sensitivity over the shift range

5.3 Spacer DBO prediction vs. expectation
The dynamic precision (3σ) observed for a 2x4 65/390 target in the case of LFLE was better than predicted (sections 4.2-4.4). Assuming the same behavior applies to SADP, precision should be of the order of ~0.05nm.

6. Model-based overlay measurement (mDBO)

The success of scatterometry for CD and profile measurement comes from the ability to model the signal formation process. The signature contains enough information that the measurement can be made by finding those parameters that give the closest fit between modeled and experimental signatures. The same approach can be applied to overlay measurement, reducing the number of measurement pads needed and providing profile data as well as overlay.

6.1 mDBO LELE sample
The first mDBO DPT structure consists of alternate photo resist and nitride lines on a silicon substrate (Fig. 16a). As mentioned in section 3, eDBO measurements are performed on four specially designed pads per direction (Fig. 4) with D designed to be around 25-35% of the pitch to ensure maximal overlay sensitivity [6]. The mDBO measurements are performed on two of the pads with shift $+D$ and $-D$. For normal incident polarized reflectometry, it is found that the TE spectrum is more sensitive to overlay than TM [6]. To reduce measurement time without compromising sensitivity, only TE spectra are collected and used for data analysis. In mDBO analysis a physical model is first set up using NanoDiffract Software to describe the sample structure. Fig. 16 shows the model of one of the pads with designed shift $+D$. Four parameters, nitride bottom CD (NI_BW), resist bottom CD (PR_BW), resist height (PR_HT) and the distance between the nitride and resist lines (S), are floated to optimize the model fit to the measured spectra. When there is an overlay error ε, the distance between the nitride and resist lines, denoted as $S(+)$ with + for positive shift, is given by $D+\varepsilon$. The model for the second pad with designed shift $-D$ is identical to the first pad except that the shift denoted by $S(-)$ is D-ε. In the regression it is assumed that the

corresponding thickness and CDs are the same for these two pads due to proximity. D is fixed to the designed value of 244nm. Fig. 18(b) displays the experimental spectrum and theoretical calculation at best fit for one of the pads. The agreement is excellent. The shape of the spectrum and fit quality for the second pad is very similar to the first one.

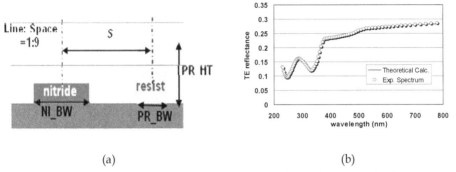

(a)	(b)

Fig. 16. (a) DPT structure with alternative photo resist and nitride lines with silicon over etch. Four parameters are floated: nitride bottom CD (NI_BW), resist bottom CD (PR_BW), resist height (PR_HT) and the shift (S) of resist from nitride lines, measured from center to center. (b) Experimental spectrum and theoretical calculation.

To check the stability and performance of the model, uncertainty and sensitivity analysis (U&SA) was performed using NanoDiffract™ software [16]. Fig. 17 shows the signal to noise ratio corresponding to a 2 nm change in the overlay error. Reasonable sensitivity is observed. The parameter correlation matrix and predicted static precision (3σ) are summarized in table 4. No strong correlation is found between overlay and other parameters. The predicted static precision for overlay is 0.16nm (3σ), which compares well with the eDBO result of 0.25nm in section 3.3.

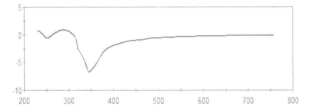

Fig. 17. Overlay signal/noise ratio. The signal corresponds to 2nm change in overlay.

Fig. 18a compares two-pad mDBO measurements with 4-pad eDBO results. Both data sets are from ~140 dies across the wafer. Excellent correlation (R² ~0.99) and a slope of 1.00 are achieved. The offset is about 0.1nm. Fig. 18b shows the histogram of the deviation of the data points from the correlation curve shown in Fig. 18a. The distribution follows a normal distribution, indicating the absence of systematic error between these two analysis methods. Standard deviation (3σ) is 1.05nm, which contains measurement uncertainties from both measurement methods.

Parameter Correlation Matrix:					Precision (3σ) (nm)
NI_BW	1				0.091
S (overlay)	-0.19	1			0.16
PR_HT	-0.47	0.54	1		0.047
PR_BW	0.05	-0.68	-0.62	1	0.052

Table 4. Parameter correlation matrix and precision predicted using model shown in Fig. 16 (a).

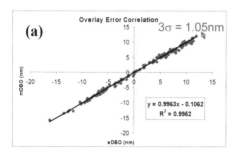
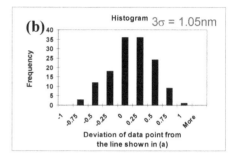

Fig. 18. (a) Correlation of mDBO and eDBO for LELE sample. (b) Histogram of the deviation of the data points form the straight line shown in (a). Data shown here for X and similar overlay error is observed for Y direction

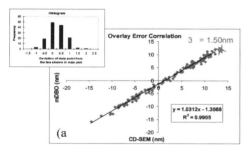
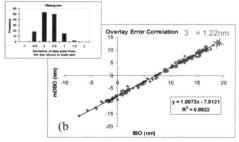

Fig. 19. (a) Correlation of mDBO and CD-SEM for LELE wafer. The inset shows the histogram of the deviation of the data points from the straight line shown in main plot. (b) Correlation of mDBO and IBO. The inset shows the histogram of the deviation of the data points from the straight line shown in main plot.

To further evaluate the accuracy of scatterometry measurement, mDBO results are compared with other metrology techniques, i.e., The CD-SEM data is from the DBO targets. Image based overlay (IBO) measurements are made on standard box-in-box targets nearby. Correlations of mDBO to these two techniques are shown in Fig. 19. A good correlation (R^2=0.99) and a slope of 1.03 are observed between eDBO and CD-SEM. The offset between eDBO and CD-SEM measurements is ~1.3nm. A good correlation (R^2=0.99) is also observed between eDBO and IBO. However, there is an offset of ~7.9 nm between the two methods.

The source of the offset is not clear. The deviation of the data points from the linear correlation curve is 1.50nm 3σ between mDBO and CD-SEM, and 1.22nm 3σ between mDBO and IBO.

6.2 mDBO LFLE sample

Targets composed of only one pad are desirable because they further reduce total target size. 2D gratings that are sensitive to overlay errors in both X and Y directions may be used [13, 17]. One example is shown in Fig. 20. A 2D lattice (similar to an IBO box-in-box target) is formed with a period on the order of hundreds of nanometers, chosen to maximize diffraction efficiency and overlay sensitivity. For IBO targets , the scale of the boxes is on the order of microns to a few tens of microns. The size of the IBO target is limited by optical resolution.

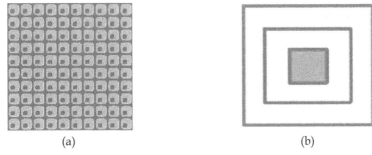

(a)   (b)

Fig. 20. (a) mDBO 2D grating target; (b) IBO box-in-box target

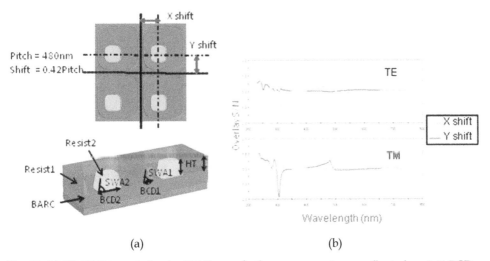

(a)   (b)

Fig. 21. (a) 2D DBO targets for the LFLE sample. Seven parameters are floated: resist1 BCD, SWA and HT, resist2 BCD, SWA and HT (coupled to resist1 HT), X shift and Y shift defined from the center of the grids (resist1) to the center of the squares (resist2). (b) Overlay S/N spectrum. The signal corresponds to 0.5nm change in X and Y overlay.

The overlay errors are extracted from the DBO target using the modeling approach. The sample structure and modeling details are shown in Fig. 21a. The pitch is 480nm and the nominal values of X shift and Y shift are 42% of the pitch. The design is symmetric in X and Y so that resist side wall angles (SWA) and bottom critical dimensions (BCD) can be coupled between X and Y directions. Seven parameters are floated: resist1 BCD, SWA and height (HT), resist2 BCD, SWA and HT (coupled to resist1 HT), X shift and Y shift defined from the center of the grids (resist1) to the center of the squares (resist2). Sensitivity analysis shows that TE spectra are more sensitive to X overlay while TM spectra are more sensitive to Y overlay (Fig. 21b). Therefore, both TE and TM spectra are used in the measurement.

The experimental spectra and RCWA fits are shown in Fig. 22. In fig. 21b, the overlay S/N corresponds to 0.5nm change in overlay. Sensitivity to overlay of the 2D targets (Fig. 11), is about half of that of 1D targets for the most sensitive wavelength, if both are normalized to 1nm. This is reasonably understood considering the reduction in the target size.

Experimental spectrum (sample) and fit (model)

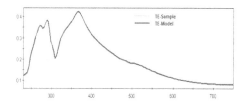 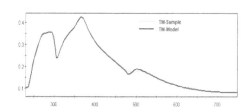

Fig. 22. Experimental TE and TM spectra and theoretical fits for the structure in Fig. 21a.

Parameters	Resist1 SWA	Resist Ht	Resist1 BCD	Y Shift	Resist2 SWA	X Shift	Resist2 BCD
Resist1 SWA	-1						
Resist Ht	-0.79	-1					
Resist1 BCD	0.95	-0.85	-1				
Y Shift	-0.03	-0.34	-0.04	-1			
Resist2 SWA	-0.62	0.28	-0.6	0.19	-1		
X Shift	-0.03	-0.37	-0.05	0.43	0.18	-1	
Resist 2 BCD	-0.5	0.001	-0.64	0.29	0.83	0.32	-1

Table 5. Parameter correlation matrix for the model shown in Fig. 21(a).

The parameter correlation matrix is shown in table 5. There are no strong correlations between overlay and the other parameters. The accuracy is first verified by measuring a series of five targets with designed shifts increasing by 2nm between two neighboring targets. The correlation of the measurement and the programmed overlay is displayed in Fig. 23a. R^2 is 0.996 and slope is 0.996. An offset of -11.36nm is observed. It comes from the local registration error due to scanner alignment errors. This can be corrected by adding an overlay offset between the layers during exposure. IBO measurements on a Blossom target next to the DBO targets show a registration error of -11.18nm, which agrees with the offset within 0.2nm. The DBO accuracy is further verified by measuring 49 fields across the wafer and correlating to Blossom measurements. The correlation plot is shown in Fig. 23b, with R^2 of ~ 0.98 and slope of ~0.97.

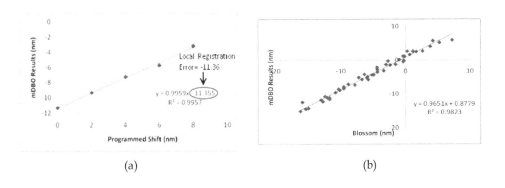

(a)   (b)

Fig. 23. (a) Correlation of mDBO 2D target measurements with programmed shifts. (b) Correlation of mDBO 2D targets with IBO blossom measurements.

6.3 mDBO LFLE performance

The dynamic repeatability (DYN 3σ), TIS mean, and TIS 3σ are reported in table 5. The dynamic repeatability is measured from 15 load/unload cycles on multiple fields (9 fields for all 1D targets, and 15 fields for the 2D targets). DYN 3σ is reported as the average of the 3σ-precisions from the measurement sites. TIS is defined as in eq. (6),

$$TIS = \frac{OVL_0 + OVL_{180}}{2}$$
(6)

where OVL_0 is the overlay measurement result at 0° loading angle, and OVL_{180} is the measurement result at 180° loading angle. The reported TIS mean is measured over 71 sites for 1D target and 49 sites for 2D targets across the wafer. For all DBO targets, TIS mean is nearly zero. TIS 3σ reported is from multiple fields (9 fields for all 1D targets, and 15 fields for the 2D targets); with OVL_0 and OVL_{180} results averaged over 15 load/unload cycles respectively. By removing the contribution from dynamic variations for each loading angle, TIS 3σ is very small (on the order of 0.01nm).

All three types of standard 1D DBO targets have shown excellent performance with TMU <0.1nm and the 2D 1x1 target has a TMU ~0.2nm (not including tool matching). It is worth mentioning that the mDBO 2x2 target has better TMU, which is a good balance between measurement performance, target size, and measurement time. Similar performance is also observed for Y.

Technology	DYNP (nm)	TIS Avg (nm)	TIS 3_σ (nm)	TMU*
Targets	X	X	X	X
eDBO 1D 2x4 target	0.088	-0.006	0.029	0.092
mDBO 1D 2x2 target	0.050	-0.006	0.028	0.058
mDBO 1D 2x1 target	0.085	0.005	0.042	0.095
mDBO 2D 1x1 target	0.172	0.057	0.120	0.209

Table 6. Performance summary of eDBO and mDBO targets (*TMU does not include tool matching)

7. Conclusion

The multi-pad empirical diffraction-based overlay (eDBO) technique is capable of controlling the overlay in double patterning optical lithography processes (DPT). The usable range of LELE DPT eDBO is ±70nm. eDBO results agree well with traditional image-based overlay (IBO) results and with overlay calculated from CD-SEM data. While good correlation and linearity between DBO and IBO was observed, a significant systematic offset can occur that appears to originate in the IBO data. Reduction in the number of pads from 2x4 to 2x3 results in only a small deterioration in precision.

DBO measurements have near-zero TIS. TMU (including tool-to-tool matching) is less than 0.5nm for both the LELE and freeze process. The overlay errors determined by eDBO (4 pad measurement) agree well with modeled DBO (2 pad measurement) data. This ability to model the signal formation process allows model-generated spectra to be used to predict measurement precision with good success.

Simultaneous model-based measurement of overlay in X&Y (2D mDBO), is possible with good results and also allows reduction in the overall target size. mDBO requires knowledge of the film stack, material optical properties and target layout and consequently more effort in creating recipes than eDBO but provides significant value in reducing measurement time and target size. In addition to overlay data, mDBO provides CD measurements and profile data for the target, which is not possible with other methods. The multi-pad DBO approach is a good method of overlay process control, especially if combined with in-chip measurements using an alternative technique [3].

8. Acknowledgement

The authors thank C. Saravanan and Nagesh Avadhany (*Nanometrics*), for their contribution to this article.

9. References

[1] C. Ludwig and S. Meyer, "Double Patterning for Memory ICs", "Lithography / Book 2/Chapter IX", ISBN 978-953-307-1356-4.

[2] ITRS, "http://www.itrs.net/Links/2009Summer/PresentationsPDF/Litho_7-2009_SF-V2.pdf"

[3] Yi-Sha Ku, Chi-Hong Tung and N. P. Smith, "In-chip overlay measurement by existing bright-field imaging optical tools," Proc. SPIE 5752, 43 (2005)

[4] P. Dasari et al., "Diffraction Based Overlay Metrology for Double Patterning Technologies," Proc SPIE 7272, 41 (2009)

[5] R. Kim et. al., "22nm half-pitch patterning by CVD spacer self alignment double patterning (SADP)," Proc SPIE 7973, 22 (2011).

[6] C. Saravanan et al., "Evaluating Diffraction Based Overlay Metrology for Double Patterning Technologies," Proc SPIE 6922, 10 (2008)

[7] J. Bishoff, R. Brunner, J. Bauer and U. Haak, "Light diffraction based overlay measurement," Proc SPIE 4344, 222 (2001)

[8] Chun-Hung Ko et al., "Comparisons of overlay measurement using conventional bright-field microscope and angular scatterometer," Proc. SPIE 5752, 987 (2005)

[9] H.-T. Huang et al., "Scatterometry-based Overlay Metrology," Proc SPIE 5038, 126 (2003)

[10] W. Yang, et al., "Novel Diffraction-based Spectroscopic Method for Overlay Metrology," Proc SPIE 5038, 200 (2003)

[11] Daniel Kandel et al., "Differential Signal Scatterometry Overlay Metrology: An Accuracy Investigation," Proc SPIE 6616, 0H1 (2007)

[12] Jie Li et al., "Advancements of Diffraction Based Overlay Metrology for Double Patterning," Proc SPIE 7971, 70 (2011)

[13] M Dusa et al., "Application of optical CD for characterization of 70mm dense lines," Proc SPIE 5752, 30 (2005)

[14] N.P. Smith et al., "Overlay metrology at the crossroads," Proc SPIE 6922, 2 (2008)

[15] Jie Li et al., "Simultaneous Overlay and CD Measurement for Double Patterning: Scatterometry and RCWA Approach," Proc SPIE 7272, 4 (2009)

[16] P Vagos et al., "Uncertainty and sensitivity analysis and its application in OCD measurements", Proc SPIE 7272, 65 (2009)

[17] B. Schultz, US Patent 7099010 "Two-dimensional structure for determining an overlay accuracy by means of scatterometry"

Part 2

Other Lithographic Technologies: Scanning Probe, Nanosphere, Inkjet Printing, etc.

6

Scanning Probe Lithography on Organic Monolayers

SunHyung Lee[1], Takahiro Ishizaki[2], Katsuya Teshima[1],
Nagahiro Saito[3] and Osamu Takai[3]
[1]Faculty of Engineering, Shinshu University,
[2]National Institute of Advanced Industrial Science and Technology (AIST),
[3]EcoTopia Science Institute, Nagoya University,
Japan

1. Introduction

Lithographic technologies for the surface modification of inorganic and organic surfaces have been developed for various devices such as sensing, data memory, single molecule electronics and biological systems. Nano and micropatterning of organic monolayers have attracted attentions for applications to biological systems in which proteins or DNA are fixed. Photolithography, microcontact printing, and electron beam lithography have usually been used as patterning techniques for organic monolayers (Hayashi et al., 2002; Hong et al., 2003; Saito et al., 2003; Hahn et al., 2004; Kidoaki & Matsuda, 1999). Although the electron-beam lithography can fabricate very small patterns, it requires an ultra-high vacuum system (Harnett et al., 2001). The resolution of photolithography is limited by the light wavelength. Moreover, these methods are based on destructive lithography, i.e., they cause damages to the organic materials.

In particular, nano-lithographic technologies have evolved in order to satisfy persistent demands for miniaturization and high-density integration of semiconductor electric circuits. Scanning probe microscopy (SPM) has been a key tool in achieving this goal. SPM can be used not only as the means that observe surface structure at sub-molecule level by a probe but also as the means that control the atomic and molecular arrangement on a substrate. As a local nano-fabrication means, the lithography technique in nanoscale range by using SPM is called to the scanning probe lithography (SPL) (Kaholek et al., 2004; Blackledge et al., 2000; Tello et al., 2002; Liu et al., 2002). In particular, a variety of lithographic techniques using SPM probe can fabricate nano-scale patterns on an organic monolayer, such as nanoshaving, nanografting, anodization SPL, dip-pen nanolithography (DPN), and electrochemical SPL. The lithography technique is used to break the material surface by using various energy sources. SPM can also be used to break the organic monolayer. For instance, nanoshaving involves mechanical scratching by physical pressure of the probe, and anodization lithography involves anodic oxidation of the substrate surface by an applied bias voltage (above 9 V) between the probe and the substrate (Jang et al., 2002; Kaholek et al., 2004; Sugimura, & Nakagiri, 1995). In the case of the anodization lithography, the oxide layer can be fabricated by anodic oxidation. As another SPL technique, the nanografting procedure combines the fabrication of a nanopattern and the binary SAM using atomic force microscopy (AFM) (Amro et al., 2000; Xu et al., 1999; Liu et al., 2000). As the other process, a probe is scanned on the matrix SAM at a high force. The matrix SAM is removed and simultaneously replaced by another molecule in the scanning area. Fig. 1 shows the nanoshaving, nanografting, and anodization lithography techniques.

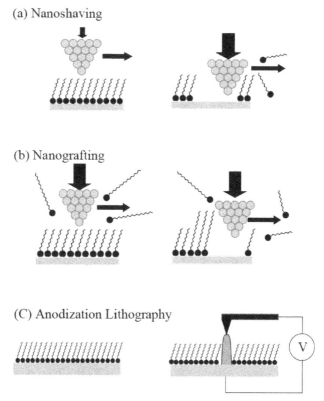

Fig. 1. Various lithographic methods using SPM probe on the organic monolayer. : (a) nanoshaving, (b) nanografting and (c) anodization lithography

In addition, dip-pen nanolithography (DPN) is a nanopatterning technique with a probe which delivers molecules to a surface via a water meniscus in the ambient atmosphere (Pena et al., 2003; Schwartz,2002; Maynor. et al., 2004). This direct-writing technique offers high-resolution patterning capabilities for a number of moleculars and biomolecular materials (ink) on a variety of substrates (paper). Nanografting and DPN are recently developed as "soft" lithographic methods, which do not cause any damage to the organic monolayers. Electrochemical SPL is also one of the soft lithographic methods. This is performed through the water column condensed between the tip of the SPM and the substrate surface. This water column can be used as a minute electrochemical cell. When a bias voltage is applied, a redox reaction occurs on the substrate surface. In the case of SAM, the functional group is converted by this redox reaction and a nanopattern is formed on the SAM surface. Fig. 2 shows the DPN and electrochemical SPL techniques.

(a) Dip-Pen Nanolithography (DPN)

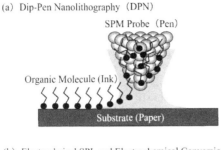

(b) Electrocheical SPL and Electrochemical Conversion Reaction

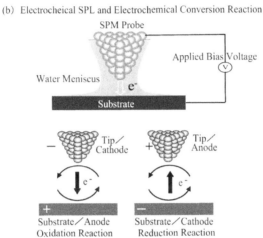

Fig. 2. Schematic illustration of dip-pen nanolithography (DPN) and electrochemical SPL.

In this chapter, we introduce anodization SPL and electrochemical SPL for fabricating the nanstcuture and controlling the function group on the various organic monolayers, such as organosilane self-assembled monolayers (SAMs) and the monolayer covalently attached to silicon through Si-C bond . First, anodization SPL techniques are expected to fabricate nanostructures on surfaces of electronic device. Their technique can remove an organic monolayer and fabricate an oxide structure by applying highly bias voltage between probe and substrate. When a high positive bias voltage is applied to the probe, an oxidation reaction proceeds and an oxide structure forms on the surface. The electrochemical reactions on SPM probe and substrate are shown in Eqs., respectively (Lee et al., 2009).

$$4H_2O + 4e^- \rightarrow 2H_2 + 4OH^- \tag{1}$$

$$M + 2H_2O \rightarrow MO_2 + 4H^+ + 4e^- \ (M : \text{substrate}) \tag{2}$$

In these reactions, electric characteristics of coated materials on the probes are important to formation of an oxide structure. The oxide structure is easily fabricated on the substrate by anodization SPL since the silicon oxide insulator layer is removed. However, the oxide structure is difficult to control and is not attained to single nanometer level though many researchers have reported on anodization SPL. We introduced fabricating technique of a three-dimensional nanostructure (nanoline structure) of silicon oxide on the hydrogen-terminated Si substrates by anodization SPL, and the effects of coated materials of SPM probe on the sizes of oxidized structures (Lee et al., 2009).

Second, we introduce a novel soft lithography based on electrochemistry through scanning probe electrochemistry for controlling the functional group on the organic monolayers (Lee et al., 2007; Lee & Ishizaki et al., 2007; Saito et al., 2005; Sugimura et al., 2004). Scanning probe electrochemistry in which materials surfaces locally oxidized or reduced by a tip of SPM is a promising technique for constructing nanostructures consisting of organic molecules. This electrochemical SPL is performed through water column which is condensed between the tip of SPM and the substrate surface. This water column can be used as a microscopic electrochemical cell. In the case of electrochemical conversion using a nanoprobe, the electrochemical reactions which proceed at the probe-sample junction are governed by the applied bias voltage and its polarity. When the substrate is polarized positively, anodic reactions, that is, oxidation reactions, proceed on its surface. On the contrary, when the substrate is polarized negatively, cathodic reactions, that is, reductive reactions, proceed. The method is expected as a key technology for future molecular nanodevices. In addition, organosilane self-assembled monolayers (SAMs) have been applied to a resist material for SPL. Here we report the chemical conversion of an organic molecular monolayer in a reversible manner using SPM. The chemical state of the monolayer from its oxidized state to reduced state or vice versa is controlled. First, an amino-terminated SAM was chemically converted into an oxidized SAM by SPL at positive-bias voltages. Moreover, this oxidized SAM was then reconverted into an amino-terminated SAM by SPL at negative bias voltages. We examine the chemical changes undergone in the scanned area from the viewpoint of surface-potential reversibility. Additionally, we introduce a electrochemical SPL to fabricate -COOH groups on an organic monolayer directly attached to silicon, which was synthesized from 1,7-octadien (OD) and hydrogen-terminated silicon. The –COOH groups on the OD-monolayer were also synthesized by the properties of surface produced by SPL.

2. Anodization scanning probe lithography on the organic monolayer on the hydrogen-terminated Si substrate

In this chapter, we introduced fabricating technique of a three-dimensional nanostructure of organic monolayer on the hydrogen-terminated Si substrates by anodization SPL. First, we fabricated the organic monolayer on the hiydrogen-termicated Si(111) wafers with an electrical resistance of 10.0-20.0 Ω-cm. Si substrates were sonicated in acetone and ethanol for 10 min, and then, cleaned by an ultraviolet (UV) light/ozone cleaning method. The substrates were exposed to vacuum UV (VUV) light (172 nm) from an excimer lamp for 30 min under atmospheric pressure and room temperature. Subsequently, the substrates were cleaned in piranha solution (H_2SO_4:H_2O_2 = 3:1) at 100 °C for 10 min and rinsed in ultrapure water. They were then etched for 15 min by immersing in 40 % aqueous ammonium fluoride solution (NH_4F). As a result, the native silicon oxide layer was removed from the substrates, and hydrogen-terminated surfaces were formed. 1-decane monolayer was prepared by a liquid phase method. The hydrogen-terminated substrates were immersed in the solution of 1-decane molecules at 150 °C for 3 h . After the immersion, the organic monolayer coated substrates were cleaned in toluene, acetone, and ethanol, and rinsed in ultrapure water.

Firstly, the 1-decane monolayer was fabricated on the hydrogen-terminated Si substrates through a liquid phase method at 150 °C. Fig. 3(a) shows the chemical structure of 1-decane molecule. The static water contact angle and film thickness of the monolayer was about 107° and 1.12 nm, respectively. A smooth surface coated with densely packed CH_3 groups (alkyl monolayer) shows a water contact angle of approximately 110°. In addition, its thickness was considered reasonable since the chain length of 1-decane molecule was estimated to be 1.32 nm. It is indicated that the 1-decane molecules formed an organic monolayer on the silicon substrate. Fig. 3(b) shows the XPS Si2p spectrum of the Si surface coated with the 1-decane monolayer. As clearly shown in Fig. 3(b), the peak of native silicon oxide was not observed at all. These results indicate that the 1-decane molecules reacted directly to the hydrogen-terminated silicon surfaces and formed the organic monolayer there. Fig. 3(c) and 3(d) show the AFM topographic and friction images of the Si surfaces after immersing in 1-decene solution. The flat terraces with steps for silicon one-atom and the changes of high friction on steps were, respectively, observed on the topographic and friction images. In the topographic image [Fig. 3(c)], the intervals of the steps and the terraces were 180 nm and 3.2 ± 0.3 Å, respectively. In addition, the difference of friction force between the flat terraces and steps was about 10 mV. These results indicated to the surface profiles of hydrogen-terminated silicon. It was therefore found that the 1-decane molecules were vertically and densely assembled, and their monolayer was formed through doubly-boned terminated group attached to the hydrogen-terminated silicon surface.

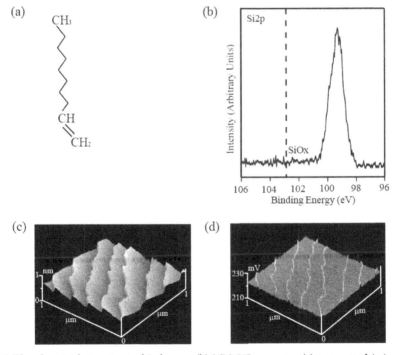

Fig. 3. (a) The chemical structure of 1-decane (b) XPS Si2p spectra (c) topographic image (d) friction image of sample surfaces after prepared 1-decane monolayer. [SH. Lee, N. Saito, O. Takai, Highly reproducible technique for three-dimensional nanostructure fabrication via anodization scanning probe lithography, Appl. Surf. Sci., 255, 7302-7306 (2009). Copyright@ELSEVIER (2009)]

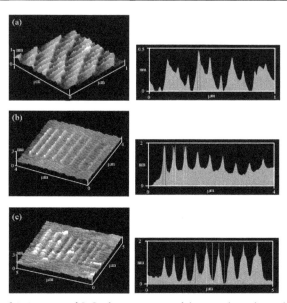

Fig. 4. The topographic images of SiOx line structure fabricated on the 1-decane monolayer and profiles for each topographic image by using various tip: (a) diamond coated Si tip, (b) Si tip and (c) Au coated Si tip. [SH. Lee, N. Saito, O. Takai, Highly reproducible technique for three-dimensional nanostructure fabrication via anodization scanning probe lithography, Appl. Surf. Sci., 255, 7302-7306 (2009). Copyright@ELSEVIER (2009)]

Next, we investigated that the effect of scanning rate on nanoline width of silicon oxide fabricated by anodization SPL. An anodization SPL was carried out on the 1-decane monolayer in air (humidity in ranging from 30 to 40 %) at the applied bias voltage of 9 V. In the anodization SPL, 1-decane monolayer was removed and the SiOx nanoline structures were formed by scanning the probes. Fig. 4(a)-4(c) show the topographic images and profiles of SiOx nanoline structure fabricated by the anodization SPL. In these experiments, various probes were used to fabricate oxide nanostructures [Fig. 4(a); diamond-coated Si probe, Fig. 4(b); Si probe (i.e., uncoated Si probe), Fig. 4(c); Au-coated Si probe]. Additionally, we investigated the effect of coated materials on the formation of the oxide nanostructure. The ranging of scanning rates for SPL is 0.1 to 5 μm/s. In these topographic images, the nanoline structure and flat terraces with steps were obviously observed under all scanning conditions. However, the widths of nanoline structures were changed with the scanning rates. Fig. 5 shows the variation in the width of nanoline structures with scanning rates and surface compositions of probes. When the diamond coated probe is used for anodization SPL, the nanoline widths were found to remain constant under all scanning conditions, and the scanning rates had no effect on the line width. The width was approximately 15 nm, which is one of the finest nanostructures in the field of SPL technique. The highly reproducible structure was fabricated by anodization SPL using the diamond-coated Si probe, and this technique is thought to be able to apply various industrial fields. On the other hand, when the Si probe (uncoated-Si) and Au-coated Si probe were used for anodization SPL, the line widths were markedly changed with the scanning rates. In the case of Au-coated Si probe, as the scanning rate increased from 0.1 to 5 μm/s, the width of line structures drastically decreased from 375 to 125 nm. For Si probe (uncoated Si probe), the line width gradually decreased with increasing scanning rate, reaching about 100 nm at 5 μm/s. The width variations were Au-coated Si probe > Si probe (uncoated) > diamond-coated Si probe. These variations of the line width could be explained the band-gap energy of the coated materials. The band-gap energies of Au, Si and diamond are respectively about 0, 1.2 and 5.47 eV. That is to say, their conductive property is thought to be greatly dependent on the line width. These results indicate that high reproducibility of oxide nanoline structures is attainable by means of anodization SPL using the diamond-coated probe or a probe which has relatively low conductivity.

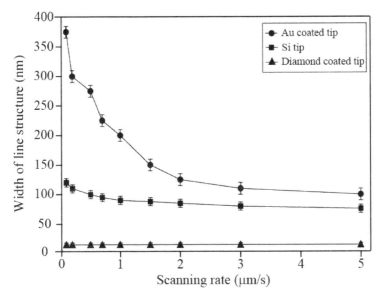

Fig. 5. The change of the line width with increasing scanning rate [SH. Lee, N. Saito, O. Takai, Highly reproducible technique for three-dimensional nanostructure fabrication via anodization scanning probe lithography, Appl. Surf. Sci., 255, 7302-7306 (2009). Copyright@ELSEVIER (2009)]

Finally, the effect of applied bias voltage on nanoline width of silicon oxide was investigated by use of various SPM probes. Fig. 6(a)-(c) show the topographic images of SiOx nanoline structures fabricated by diamond-coated Si, Si (uncoated) and Au-coated Si probes, respectively. The applied bias voltage and scanning rate were, respectively, fixed at 7 V and 1 μm/s. The obtained widths were 15, 60 and 100 nm in Fig. 6(a)-(c), respectively. The line width fabricated at the bias voltage of 7 V decreased compared to that at 9 V when Au-coated Si and uncoated Si probes were used. For the diamond-coated probe, the nanoline structure maintained the width of 15nm, that is, the applied bias voltages had no effect on the nanoline width. These results are also attributed to the band-gap energy of the coated materials. In particular, three-dimensional oxide nanostructures fabricated by the diamond-coated probe showed highly reproducibility even though various scanning rates and applied bias voltages are used in aondization SPL. Therefore, this technique is expectable to be applied to fabricate a wide variety of nanodevices, that have three-dimensional structures, in various industrial fields.

(a)

(b)

(c)

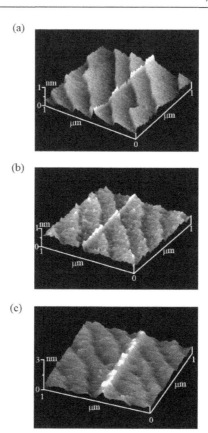

Fig. 6. Topographic images of line structure fabricated by (a) diamond coated Si tip, (b) Si tip and (c) Au coated Si tip, respectively. The applied bias voltage and scanning rate are 7 V and 1μm/s, respectively. [SH. Lee, N. Saito, O. Takai, Highly reproducible technique for three-dimensional nanostructure fabrication via anodization scanning probe lithography, Appl. Surf. Sci., 255, 7302-7306 (2009). Copyright@ELSEVIER (2009)]

3. Electrochemical scanning probe lithography on organic monolayer

A future tool of nanolithography may be nanopattern drawing based on SPM. Several reports describe SPL in which a sample surface is chemically converted or mechanically deformed at the probe-sample junction. The characteristic feature of all these methods is that they allow permanent nanopatterns to be drawn on the substrate. Scanning probe electrochemistry in which materials surfaces locally oxidized or reduced by a tip of SPM is a promising technique for constructing nanostructures consisting of organic molecules. This electrochemical SPL is performed through water column which is condensed between the tip of SPM and the substrate surface. This water column can be used as a microscopic electrochemical cell. When the bias voltage is applied, redox reaction proceeds on the substrate surface. The method is expected as a key technology for future molecular nanodevices. In addition, organosilane SAMs have been applied to a resist material for SPL.

4. Nano-probe electrochemistry on amino-terminated self-assembled monolayers toward nano memory

In this chapter, we investigated that an amino-terminated SAM was electrochemically converted into an oxidized SAM by SPL at positive bias voltages. Moreover, this oxidized SAM was then reconverted into an amino-terminated SAM by SPL at negative bias voltages. The chemical conversions of amino groups were confirmed by Kelvin probe force microscopy (KPFM), atomic force microscopy (AFM) and the site-selective adsorption of carboxyl-modified fluorescent spheres. We examined the chemical changes undergone in the scanned area from the viewpoint of surface-potential reversibility. Fig. 7 schematically illustrates of the experimental procedure. An amino-terminated SAM was prepared by chemical vapor deposition (CVD) from *p*-aminophenyltrimethoxysilane (APhS) on n-type silicon (111) wafers with electrical resistivity of 4-6 Ω/cm. First, the silicon substrate was cleaned in acetone, ethanol, and deionized water, in that order. After cleaning, the silicon substrate was irradiated by 172 nm vacuum ultraviolet light in air for 20 min. This removed organic contaminants and introduced silanol groups onto the substrate surface. Next, each cleaned silicon substrate was placed together with a glass vessel filled with APhS liquid in a Teflon container. The Teflon container was sealed and placed in an oven with the temperature kept at 100 °C. The reaction time was 1 h. The heated APhS liquid vaporized and hydrolyzed. The hydrolyzed APhS reacted with the silanol groups on the silicon substrate resulting in the fabrication of an amino-terminated SAM.

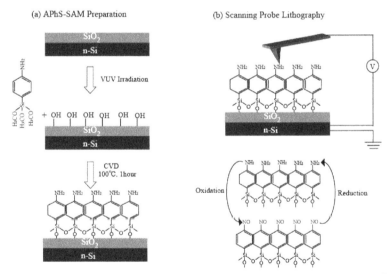

Fig. 7. Preparation and electrochemical scanning probe lithography of APhS-SAM [N. Saito, SH. Lee, T. Ishizaki, J. Hieda, H. Sugimura, O. Takai, Surface potential reversibility of an amino-terminated self-assembled monolayer based on nanoprobe chemistry, J. Phys. Chem. B, 109, 11602-11605 (2005) . Copyright@American Chemical Society (2005)]

We fabricated the APhS-SAM through the CVD method. The formation of APhS SAM was confirmed by the ellipsometry, water contact angle and X-ray photoelectron spectroscopy (XPS) measurement. In our SPL system, electrons were transferred between a gold-coated probe and the silicon substrate through the APhS SAM and an adsorbed water layer. The adsorbed water played the same role as water in a typical, macroscopic electrochemical cell system. The electrochemical conversion was conducted with a gold-coated silicon nanoprobe with a force constant of 1.8 N/m and a resonance frequency of 23.2 kHz. The probe was scanned in air at a relative humidity of 35% to 40%.

Firstly, the formation of APhS SAM was confirmed by the ellipsometry, water contact angle and XPS measurement. The static water contact angle of the sample surface after preparation was about 60°. Its film thickness obtained by ellipsometry was ca. 0.58 nm. This thickness was considered reasonable since the chain length of the APhS molecule was estimated to be 0.6 nm by semiempirical molecular orbital calculation using the AM1 Hamiltonian. The XPS N1s spectrum of the sample seen in Fig. 8 shows that the N1s binding energy is 399.6 eV. This binding energy agrees approximately with a previously reported value (400.0 eV). These results indicated the formation of APhS SAM on the SiO_2 substrate.

To show the effect of adsorbed water, the nano-probe was scanned at the pressure of 10^{-6} Pa. Fig. 9 shows topographic and surface-potential images scanned at 2 and 5 V in both air and vacuum. On the basis of these images, it is obvious that the chemical reaction does not proceed in vacuum. Thus, adsorbed water is considered necessary for the chemical conversion.

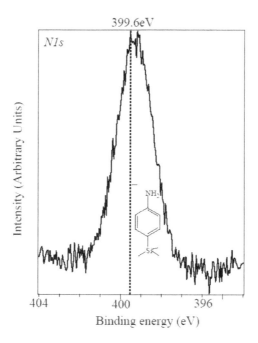

Fig. 8. XPS *N1s* spectrum of the Si surface covered with APhS-SAM.

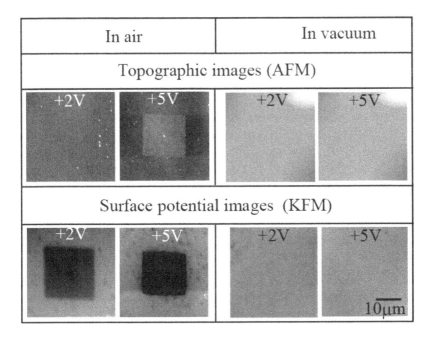

Fig. 9. Topographic and surface-potential images scanned at +2 and +5 V in air and vacuum. [N. Saito, SH. Lee, T. Ishizaki, J. Hieda, H. Sugimura, O. Takai, Surface potential reversibility of an amino-terminated self-assembled monolayer based on nanoprobe chemistry, J. Phys. Chem. B, 109, 11602-11605 (2005) . Copyright@American Chemical Society (2005)]

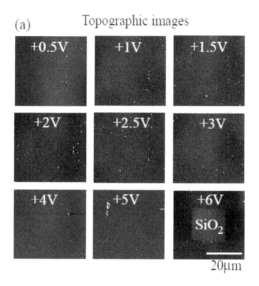

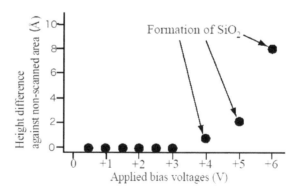

Fig. 10. (a) AFM topographic images and (b) the height difference against the non-scanned area after scanning at positive bias voltages.

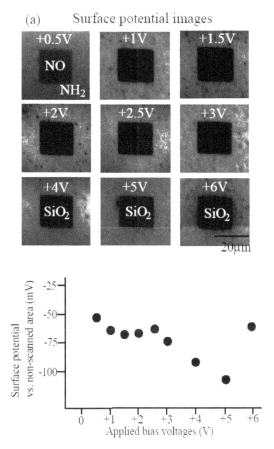

Fig. 11. (a) Surface potential images obtained by KFM and (b) the change in surface potential on the area scanned at positive bias voltages.

The nanoprobe was scanned across the sample surface over an area of 20μm X 20μm at bias voltages of 0.5-6 V. Fig. 10 shows both AFM topographic images and the height difference against the nonlithographic area after scanning. A slight protuberance can be observed in the topographic images of samples scanned at the bias voltages of 4 to 6 V. This protuberance became much higher at the bias voltages of 4 to 6V. These protuberances are due to the production of SiO_2, which resulted from the decomposition of the as-deposited APhS SAM and the oxidation of silicon. This demonstrates that there is no possibility of chemical reversibility at these bias voltages since the SAM has been damaged. Specifically, there is no possibility of surface-potential reversibility. On the other hand, no change can be observed in the topographic images of samples scanned at the bias voltages of 0.5-3 V. At these voltages, it is possible that the framework of the SAM molecules remained intact. Fig. 11 shows both surface-potential images obtained by KFM and the change in surface potential on the scanned region. The surface potential shifted negatively, which can be roughly explained by the apparent dipole moment of the SAM. The apparent dipole moment of the untreated APhS SAM is in the direction from the sample surface to the substrate. This can explain the negative shifts of the surface potential in the scanned area. In addition, amino surfaces were converted into nitroso surface at the bias voltages of 1 V to 3 V because surface potential contrast was nearly constant. In surface potential image, the surface potential reversed with the applied bias voltage.

To confirm surface-potential reversibility, a nanoprobe scanning series was performed as follows. At first, (a) a 60 µm × 60 µm square region was oxidized, and (b) a 20 µm × 20 µm square region in the 60 µm × 60 µm square region was reduced. The scan rates were 0.5 and 1.0 Hz, respectively. Fig. 12 shows illustrates of these (a) experimental processes and (b) surface potential image of scanned area. In surface potential image, the surface potential reversed with the applied bias voltage. Fig. 13 shows (a) schematic illustration of selective adsorption of carboxyl- modified fluorescent spheres after AFM lithography and (b) dark field image acquired by optical microscope after immersion of the sample in Fig. 12 in a pH 4 solution containing carboxyl-modified fluorescent spheres. The -COOH and -NH₂ groups in the pH 4 solution were converted into -COO⁻ and -NH₃⁺ ion groups. Thus, the selective adsorption of fluorescent spheres onto the NH₂ regions proceeded due to attractive electrostatic interaction. In the pH 4 solution, amino-modified fluorescent spheres were repulsed in regions with –NO terminated groups. Therefore, the bright and dark areas correspond to the NH₂ and NO surface, respectively. These indicated that the nitroso terminated surfaces were reconverted into amino terminated surfaces with a negative bias voltage.

(a)

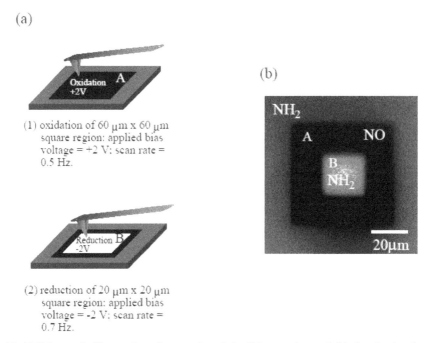

(1) oxidation of 60 µm x 60 µm square region: applied bias voltage = +2 V; scan rate = 0.5 Hz.

(b)

(2) reduction of 20 µm x 20 µm square region: applied bias voltage = -2 V; scan rate = 0.7 Hz.

Fig. 12. (a) Schematic illustrations for a series of the lithography and (b) the obtained surface potential image.

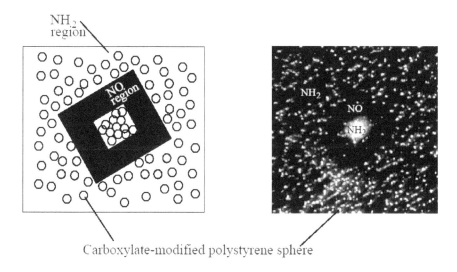

Fig. 13. (a) Schematic illustration of selective adsorption of carboxylate-modified polystyrene spheres after AFM lithography and (b) dark field image acquired by optical microscope after immersion in a pH 4 solution containing carboxylate-modified polystyrene spheres, followed by successive scanning probe lithography.

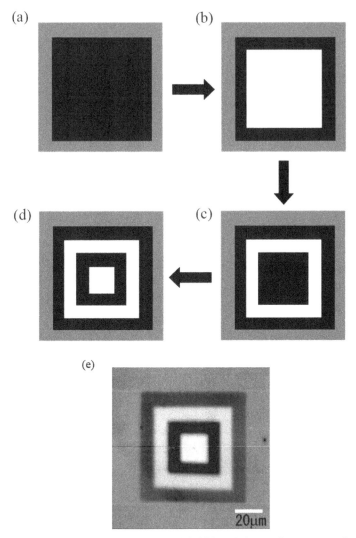

Fig. 14. Illustrations of reversibility processes ((a)-(d)) and the surface potential image of multu-scanned area (e).

In addition, we investigated the multi-reversible conversion of the APhS surface. Firstly, (a) a 80 μm × 80 μm square region was oxidized, and (b) a 60 μm × 60 μm square region in the 80 μm × 80 μm square region was reduced. Moreover, (c) the 40 μm × 40 μm square region in the 60 μm × 60 μm square region was oxidized, and then (d) the 20 μm × 20 μm square region the 40 μm × 40 μm square region was reduced. The scan rates were 0.5, 0.7, 1.0, and 2.0 Hz, respectively. Fig. 14 (a) and (b) show illustrations of these experimental processes and the surface potential image of the scanned area, respectively. In the surface potential image, the surface potential reversed with the applied bias voltage. These indicated the multi-reversible conversion of amino terminated surfaces. Thus, we can control the surface-potential reversibility on the amino-terminated SAM by controlling the applied bias voltage. This reaction formula is as follows .

$$-NH_2 + H_2O \leftrightarrow -NO + 4H^+ + 4e^- \tag{3}$$

Finally, by making use of this phenomenon of surface-potential reversibility, we demonstrated surface-potential memory. First, a square region 10 μm × 10 μm was oxidized at the bias voltage of 2 V. Next, dotted areas in the oxidized region were selectively reduced at the bias voltage of -2 V. Finally, the 10 μm × 10 μm square region was again oxidized at the bias voltage of 2 V. Fig. 15 shows the surface potential changes corresponding to (a) "writing" and (b) "erasing" with the experimental process. In Fig. 15 (a), sixteen bright areas corresponding to the chemically converted region can be observed. These sixteen areas disappeared after "erasing," as shown in Fig. 15 (b). Although this "surface potential memory" has not yet been highly integrated, it has the potential to perform as ultra-integrated memory.

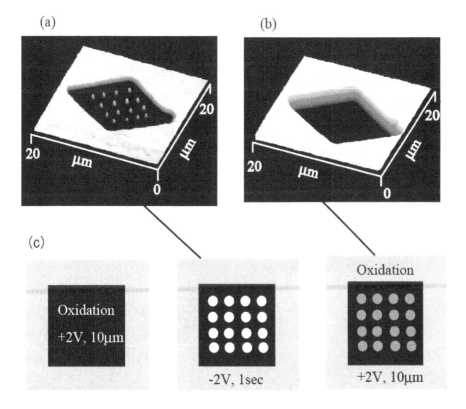

Fig. 15. Surface-potential memory: (a) "writing" state, (b) "erasing" state (c) and the experimental process.

4.1 Electrochemical lithography of 1,7-octadiene monolayers covalently linked to hydrogen-terminated silicon using scanning probe microsoopy

The organic monolayer covalently attached to silicon through Si-C bond has been expected to have better chemical resistivity compared to organosilane monolayers. The Si-C interface provides a good electronic property for molecular devices constructed on the silicon. In particular, the construction of hybrid organic- molecule/silicon devices is a promising approach for the future molecular devices. To realize such devices, it is vital to establish the fabrication technology for microstructure of the organic monolayer. The electrochemical SPL is performed through the water column condensed between the tip of SPM and the substrate surface. This water column can be used as a minute electrochemical cell. When the bias voltage is applied, a redox reaction proceeds on the substrate surface. Through reversible chemical SPL, we successfully controlled this redox reaction so that an NH_2-terminated organosilane monolayer surface was converted into an NO-terminated surface. However, this organosilane monolayer suffered from electrical defects due to the presence of SiOx. A 1,7-octadiene (OD) monolayer was directly formed on a hydrogen-terminated silicon surface by radical reaction. In this section, we report the electrochemical conversion of the vinyl-terminated groups on an OD monolayer induced by a nanoprobe.

Si (111) with electrical resistivity of 10.0-20.0 Ω/cm was used as for the substrates. Fig. 16 shows schematic illustrations of the experimental process. The substrates were cleaned in piranha solution ($H_2SO_4 : H_2O_2 = 3 : 1$) at 100 °C for 10 min and rinsed in ultrapure water. They were then etched for 15 min by immersing in 40 % aqueous ammonium fluoride solution (NH_4F). The surface was hydrogen-terminated and silicon oxide was removed by this immersion. The OD monolayer was prepared by the liquid phase method. The substrates were immersed in OD solution at 120 °C for 1h. After immersion, the samples were cleaned in toluene, acetone, ethanol and rinsed in ultrapure water.

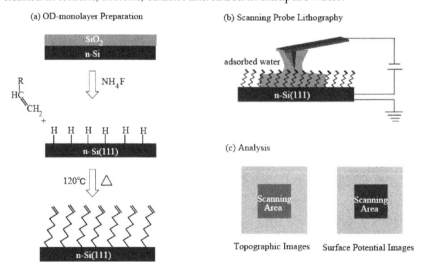

Fig. 16. Schematic illustrations of the experimental process: (a) preparation of the OD-monolayer, (b) electrochemical scanning probe lithography and (c) chemical conversion analysis. [SH. Lee, T. Ishizaki, N. Saito, O. Takai, Electrochemical Soft Lithography of an 1,7-octadiene Monolayer Covalently Linked to Hydrogen-Terminated Silicon using Scanning Probe Microscope, Surf. Sci., 601, 4206-4211 (2007). Copyright@ELSEVIER (2007)]

Fig. 17 (a) shows the XPS Si *2p* spectra of silicon substrate surfaces before and after the immersion in 40 % aqueous ammonium fluoride solutions (NH₄F). In the spectrum of the sample surface before the immersion, the peak at 104 eV attributed to SiO_2 was observed, while no appreciable peak related to the oxide was observed in the spectrum after the immersion. This indicates that the native oxide layer on the silicon substrate is completely removed after the immersion. Fig. 17 (b) show the AFM topographic image of the silicon substrate surface. The AFM image has flat terraces with the steps for silicon one-atom. The distance and the height difference between steps were evaluated to be 180 nm and 0.32±0.03 nm, respectively, as shown in Fig. 17 (b). These results reveal that the silicon surfaces are terminated with hydrogen. In order to deposit the OD monolayers, the substrate was immersed in the OD solution heated at 120 °C for 1 h. After the immersion, the water contact angle of the OD monolayers became saturated at approximately 80° at the reaction time of 1 hour. Furthermore, the film thickness of 1.2 nm corresponded approximately to the distance from Si to -CH₂ end groups in the precursor, as determined by ellipsometry. Fig. 17 (c) shows the topographic image of the OD monolayer surface. The distance and the height difference between steps, as shown in Fig. 17 (c), were 120 nm and 0.28±0.03 nm, respectively. These values are well in agreement with that of hydrogen-terminated silicon surface. The OD monolayers were stably attached to the hydrogen-terminated Si surfaces, since the parallel monoatomic steps were observed on the OD molecule. This means that the OD molecule was deposited at a monolayer on the substrate. These surfaces are very stable and can be stored for several weeks without any change in the topographic properties.

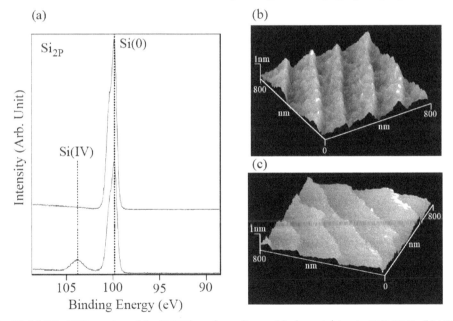

Fig. 17. (a) Si2p XPS spectra of the Si(111) surface after and before etching in 40% NH4, (b)AFM image (800nm×800nm)of a H-terminated Si(111) surface, (c) AFM image (800nm×800nm) of a OD-monolayer surface [SH. Lee, T. Ishizaki, N. Saito, O. Takai, Local Generation of Carboxyl Groups on an Organic Monolayer through Chemical conversion using Scanning Probe Anodization: Mater. Sci. Eng. C, 27, 1241-1246 (2007) Copyright@ELSEVIER (2007)]

The OD monolayers were selectively oxidized in air through the photomask by vacuum ultraviolet (VUV) light irradiation for 10 min. The areas irradiated were converted into -COOH groups due to photochemical oxidation, thus dividing the small surface into distinct -CH$_2$ and -COOH end groups regions. Fig. 18 (a) shows the XPS C1s spectra of OD monolayer surfaces before and after the irradiation. Its peak of 289.6 eV is assigned to the carboxyl group. Fig. 18 (b) shows the surface potential image (KFM) of the OD monolayer. The dark and bright regions correspond to the CPD images of low and high surface potential, respectively. In this figure, the surface potential for the irradiated OD monolayer surfaces was 20 mV lower than that of the unirradiated surfaces. The change of the surface potential indicates that -CH$_2$ end groups on the OD monolayer were chemically converted into -COOH end groups. The end groups of the OD monolayers were confirmed by the selective adsorption of amino-modified fluorescence spheres in a pH 4 solution. The -COOH and -NH$_2$ groups in the pH 4 solution were converted into -COO- and -NH$_3^+$ ion groups, so that the selective adsorption of fluorescence spheres on to the substrate proceeded due to their attractive interaction to the surface. Under this pH condition, the regions of –CH$_2$ end groups on the surface were not negatively charged and the amino-modified polystyrene fluorescence spheres did not adsorb onto it. Fig. 18 (c) shows an image acquired by dark field microscopy of the micropatterned CH$_2$ / COOH sample after immersion. The lighter areas between the dark rectangular regions correspond to the COOH terminated regions. This dark-field image indicates that amino-modified polystyrene fluorescence spheres selectively adsorbed on the -COOH end group regions since scattered light due to surface roughness can be observed. From these results, we determined that an -CH$_2$ end group had been successfully converted into the COOH terminated surface through chemical lithography, i.e., photolithography.

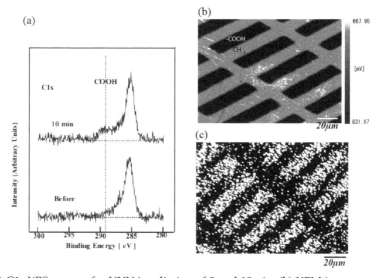

Fig. 18. (a) C1s XPS spectra for VUV irradiation of 0 and 10min, (b) KFM image (150µm×150µm) of OD-monolayer irradiated for 10min, (c) optical microscope image of the irradiated OD-monolayer which adsorbed amino terminated particles.[SH. Lee, T. Ishizaki, N. Saito, O. Takai, Local Generation of Carboxyl Groups on an Organic Monolayer through Chemical conversion using Scanning Probe Anodization: Mater. Sci. Eng. C, 27, 1241-1246 (2007). Copyright@ELSEVIER (2007)]

A gold-coated probe was scanned in air on OD monolayers at the bias voltage of 1 V and the scanning rate of 2 Hz. The areas scanned were a square, 20 μm on a side. Fig. 19 (a) and (b) shows the topographic and the surface potential images, respectively. The surface potential for the area scanned was 20 mV more negative than that of the area non-scanned and no change was observed in the topographic images. The value of surface potential difference between the scanned and non-scanned areas by SPL is well in agreement with that by photo oxidation, i.e., VUV light. These results indicate that the scanned OD monolayer surfaces are completely oxidized and chemically converted into COOH terminated surfaces without decomposition of the OD monolayer and siloxane networks. In order to verify that the chemical conversion with SPL was electrochemically proceeded on the areas scanned through the water column, the SPL was carried out in vacuum. The conditions were as follows: the bias voltage : 1 V, the pressure : 10^{-4} Pa, the scanned rate : 2 Hz, and the scanned areas : a square of 20 μm on a side. Fig. 20 shows the topographic and the surface potential images on the scanned area. In both images, no change was observed. This means that the chemical conversion on the scanned areas does not proceed under such a condition. Therefore, we can conclude that the chemical conversions with SPL are based on electrochemical reactions through the water column.

Electrochemical SPL was performed on the OD monolayer at the scanning rate of 2 Hz and at bias voltages of -3 V to 3 V. Fig. 21 shows representative surface potential images, topographic images, the changes in surface potential, and the height difference against the non- lithographic regions. In topographic images, no change in height difference was observed at any of the bias voltages. This indicates that the scanning caused no change in microscopic morphology under these conditions. On the other hand, the surface potential changed remarkably. The dark and bright regions in the surface potential images correspond to low and high surface potential regions, respectively. When the bias voltage was positively applied, oxidation proceeded on the substrate. At the positive bias voltage, the scanned regions were oxidized and showed lower surface potential than the unscanned areas. The surface potential contrast was nearly constant at approximately -18 mV at bias voltages of 1 V to 2 V. However, the surface potential contrast at bias voltages of 2 V to 3 V gradually decreased in proportion to voltage evolution.

On the other hand, the surface potential contrast at negative bias voltages gradually increased in proportion to voltage evolution. The surface potential contrast at bias voltages of 0 V to -1 V was negative, indicating that some oxidation proceeded under these conditions. This was due to the difference in contact potential between the Au-coated probe and the Si substrate. The surface potential constant was positive at bias voltages from -1.5 V to -3 V. This change in surface potential indicates that reduction reactions occurred on the substrate surface due to the applied negative bias voltage. Considering these AFM and KPFM results, the electrochemical conversion of the vinyl-terminated groups is believed to have been governed by the applied bias voltage.

Using XPS, we investigated the conversion of vinyl terminated groups at each bias voltage. Fig. 22 (a) and (b) show XPS Si2P and C1s spectra for sample surfaces after probe-scanning at bias voltages of 1 V and 3 V. The C1s spectrum in Fig. 22 (b) shows an additional peak from -COOH groups at 288.5 eV at the bias voltage of 1 V, but not at 3 V. In addition, a silicon oxide peak at 103.4 eV can be seen in Fig. 22 (a) at the bias voltage of 3 V, but not at 1 V. The intensity of the alkyl chain peak in Fig. 22 (b) at the bias voltage of 3 V was observed to be lower than that at 1 V. These XPS results indicate that the vinyl functional groups were oxidized and converted into carboxyl groups at the bias voltage of 1 V. The surface potential of the carboxyl surface was lower than that of the vinyl-terminated surface because the carboxyl groups had more negative dipole moment. This agrees with the KFM results. In addition, OD molecules on the sample surface were decomposed and silicon oxide was formed at the bias voltage of 3 V. However, the fact that there was no change of AFM morphology at the bias voltage of 3 V was probably due to the formation of "depthless" silicon oxide, that is, the partial decomposition of OD molecules, and to the effect of absorbed water on the sample surface.

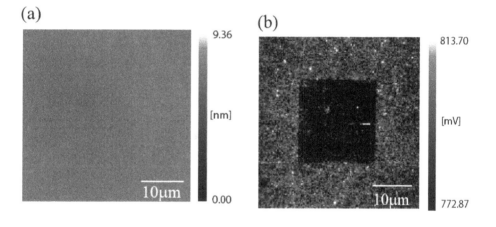

Fig. 19. (a) Topographic images and (b) surface potential image after the scanning at the rate of 2Hz. [SH. Lee, T. Ishizaki, N. Saito, O. Takai, Local Generation of Carboxyl Groups on an Organic Monolayer through Chemical conversion using Scanning Probe Anodization: Mater. Sci. Eng. C, 27, 1241-1246 (2007). Copyright@ELSEVIER (2007)]

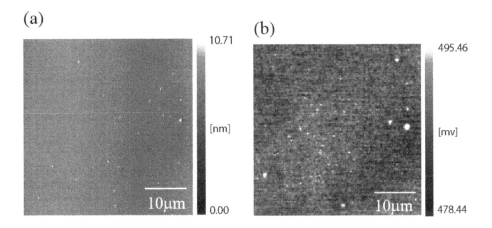

Fig. 20. (a) Topographic image and (b) surface potential image after the scanning in vacuum.[SH. Lee, T. Ishizaki, N. Saito, O. Takai, Local Generation of Carboxyl Groups on an Organic Monolayer through Chemical conversion using Scanning Probe Anodization: Mater. Sci. Eng. C, 27, 1241-1246 (2007). Copyright@ELSEVIER (2007)]

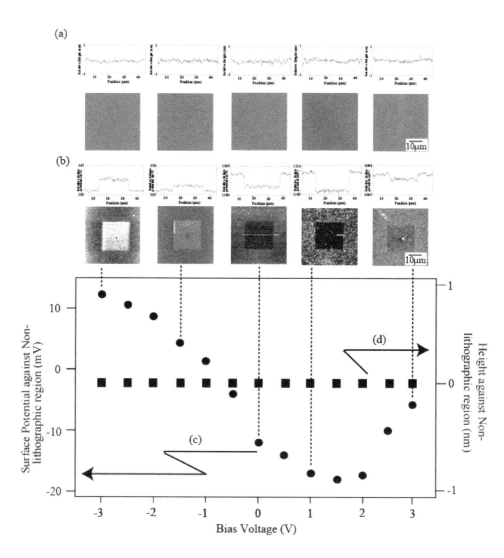

Fig. 21. (a) Representative topographic images, (b) representative surface potential images, (c) change of surface potential against non-lithographic regions; and (d) height difference against non-lithographic regions. Electrochemical SPL was performed at bias voltages of -3V to 3V. [SH. Lee, T. Ishizaki, N. Saito, O. Takai, Electrochemical Soft Lithography of an 1,7-octadiene Monolayer Covalently Linked to Hydrogen-Terminated Silicon using Scanning Probe Microscope, Surf. Sci., 601, 4206-4211 (2007). Copyright@ELSEVIER (2007)]

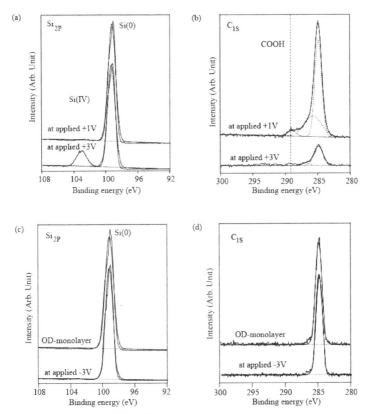

Fig. 22. (a) XPS Si2P spectra and (b) XPS C1s spectra of sample surfaces after probe-scanning at bias voltages of 1V and 3V. (c) XPS Si2P spectra and (d) XPS C1s spectra of an OD-monolayer surface and a sample surface after probe-scanning at the bias voltage of -3V. [SH. Lee, T. Ishizaki, N. Saito, O. Takai, Electrochemical Soft Lithography of an 1,7-octadiene Monolayer Covalently Linked to Hydrogen-Terminated Silicon using Scanning Probe Microscope, Surf. Sci., 601, 4206-4211 (2007). Copyright@ELSEVIER (2007)]

Fig. 22 (c) and (d) show XPS Si2P and C1s spectra of an unscanned OD-monolayer surface and a sample surface scanned at the bias voltage of -3 V. In the C1s XPS spectrum in Fig. 22 (d), the intensity of the alkyl chain peak for the surface scanned at the bias voltage of -3 V was the same as that for the OD monolayer. The silicon oxide peaks in the Si2p spectra were not observed in all cases [Fig. 22 (a) and (c)]. These XPS results indicate that OD molecules were not decomposed at negative bias voltage. We believe that the vinyl functional groups were reduced and converted into cyclobutane rings. No peak for these cyclobutane rings was observed in the C1s spectra since such a peak is generally weak. However, in view of the AFM and KFM results, we consider that cyclobutane rings form in the same manner as they are known to in photochemical and thermal reactions.

Fig. 23 shows a schematic illustration of the mechanism of electrochemical SPL on the OD monolayer. In this Section, two factors were considered: the alkyl radical reaction from frictional heat due to the probe scanning, and the redox reaction on the sample surface caused by polarization due to the applied bias voltage. First, alkyl radicals were formed by frictional heat. Next, the redox reaction occurred on the sample surface in the radical atmosphere. With positive bias voltages, the oxidation reaction easily occurred on the sample surface due to polarization in adsorbed water. Thus, the conversions on the sample surface were governed by the oxidation reaction. The vinyl-terminated groups of the OD monolayer were converted into carboxyl groups at positive bias voltage. However, the reduction reaction on the sample surface rarely occurred at negative bias voltages because the dissolved oxygen was preferentially reduced in adsorbed water. Thus, in this case, the surface reaction was governed by alkyl radicals. The formation of cyclobutane rings was considered to have occurred due to alkyl radical combinations at the negative bias voltage.

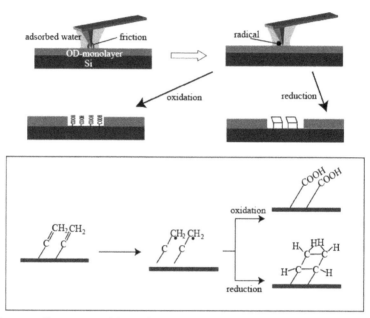

Fig. 23. Schematic illustration of the redox reaction induced by electrochemical SPL [SH. Lee, T. Ishizaki, N. Saito, O. Takai, Electrochemical Soft Lithography of an 1,7-octadiene Monolayer Covalently Linked to Hydrogen-Terminated Silicon using Scanning Probe Microscope, Surf. Sci., 601, 4206-4211 (2007). Copyright@ELSEVIER (2007)].

In support of the SPM and XPS results, the oxidized groups of the OD monolayer were confirmed by the selective adsorption of amino-modified fluorescent spheres. The -COOH and $-NH_2$ groups in the pH 4 solution were converted into $-COO^-$ and $-NH_3^+$ ion groups. Thus, the selective adsorption of fluorescent spheres onto the COOH regions proceeded due to attractive electrostatic interaction. In the pH 4 solution, regions with vinyl terminated groups were not negatively charged, and the amino-modified polystyrene fluorescent spheres were repulsed. Fig. 24 (a) shows the mechanism of this selective adsorption of the amino-modified fluorescence spheres. Fig. 24 (b) shows a dark field image of samples scanned at the bias voltage of 1 V after immersion in the solution of amino-modified fluorescent spheres. The bright areas correspond to the areas scanned at the bias voltage of 1 V which site-selectively adsorbed the fluorescent spheres. This confirms that vinyl terminated groups of the OD monolayer were converted into COOH terminated groups by scanning at the applied bias voltage of 1 V.

(a)   (b)

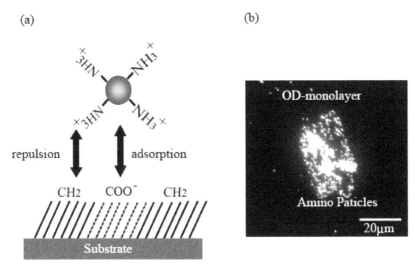

Fig. 24. (a) Selective adsorption of amino-modified fluorescent spheres on a surface scanned at the bias voltage of 1V; (b) dark field image of a patterned surface after immersion in a pH 4 solution containing amino-modified fluorescent spheres [SH. Lee, T. Ishizaki, N. Saito, O. Takai, Electrochemical Soft Lithography of an 1,7-octadiene Monolayer Covalently Linked to Hydrogen-Terminated Silicon using Scanning Probe Microscope, Surf. Sci., 601, 4206-4211 (2007). Copyright@ELSEVIER (2007)].

5. Conclusion

In this chapter, we introduced the chemical conversion of functional groups on the organic monolayer by electrochemical SPL. The three-dimensional nanostructures of silicon oxide were successfully fabricated by decomposing the 1-decane monolayer and subsequent oxidizing the hydrogen-terminated Si surfaces via anodization SPL. The size and reproducibility of oxide nanoline structures were greatly dependent on the sorts of probes for anodization SPL. In the case of Au-coated Si and uncoated Si probes, the obtained nanoline structures were changed with the scanning rates and the applied bias voltages. On the other hand, the nanotexture fabrication using the diamond-coated probe showed one of the finest structures (15 nm nanoline) and highly reproducibility even though any fabrication conditions such as scanning rate and applied bias voltage are used in the anodization SPL.The amino surface on SAM was oxidized and converted into a nitroso surface at bias voltages of 0.5 to 3 V. The functional groups on APhS SAM were reversibly converted by controlling the applied bias voltage. It was also demonstrated that the surface-potential memory was based on surface potential reversibility. In addition, the vinyl-terminated groups of the OD monolayer were site-selectively oxidized and chemically converted into carboxyl groups at bias voltages of 1 to 2 V. OD molecules on the sample surface were decomposed and silicon oxide was formed at bias voltages greater than 3 V. On the other hand, CH_2-terminal groups were converted into cyclobutane rings at bias voltages of less than −1.5 V. Recently, the research into applications of SAMs has progressed rapidly because of their ability to modify surfaces functionally and provide hydrophobicity, hydrophilicity, or biocompatibility. However, the reproducibility of the formation of SAMs is difficult. In this chapter, the formation mechanism of SAMs has been studied for high reproducibility. The control of surface properties by fabrication of micro/nanosized domains composed of SAMs is expected to be applied in the field of biomaterials. In addition, electrochemical SPL is also expected to be applied to various devices.

6. References

Hayashi, K.; Saito, N.; Sugimura, H.; Takai, O. & Nagagiro, N. (2002). Regulation of the Surface Potential of Silicon Substrates in Micrometer Scale with Organosilane Self-Assembled Monolayers. *Langmuir*, Vol.8, No.20 (September 2002), pp. 7469-7472, ISSN 0743-7463.

Hong, L.; Sugimura, H.; Furukawa, T. & Takai, O. (2003). Photoreactivity of Alkylsilane Self-Assembled Monolayers on Silicon Surfaces and Its Application to Preparing Micropatterned Ternary Monolayers. *Langmuir*, Vol.19, No.6 (February 2003), pp. 1966-1069, ISSN 0743-7463.

Saito, N.; Wu, Y.; Hayashi, K.; Sugimura, H. & Takai,O. (2003). Principle in Imaging Contrast in Scanning Electron Microscopy for Binary Microstructures Composed of Organosilane Self-Assembled Monolayers. *The Journal of Physical Chemistry B*, Vol.107, No.3 (December 2002), pp. 664–667, ISSN 1089-5647.

Hahn, J. & Webber, S. E. (2004). Graphoepitaxial Deposition of Cationic Polymer Micelles on Patterned SiO_2 Surfaces. *Langmuir*, Vol.20, No.4 (January 2004), pp. 1489-1494, ISSN 0743-7463.

Kidoaki, S. & Matsuda, T. (1999). Adhesion Forces of the Blood Plasma Proteins on Self-Assembled Monolayer Surfaces of Alkanethiolates with Different Functional Groups Measured by an Atomic Force Microscope. *Langmuir*, Vol.15, No.22 (September 1999), pp. 7639-7646, ISSN 0743-7463.

Harnett, C. K.; Satyalakshmi, K. M. & Craighead, H. G. (2001). Bioactive Templates Fabricated by Low-Energy Electron Beam Lithography of Self-Assembled Monolayers. *Langmuir*, Vol.17, No.1 (December 2000), pp. 178-182, ISSN 0743-7463.

Kaholek, M.; Lee, W. K.; LaMattina, B.; Caster, K. C. & Zauscher, S. (2004). Fabrication of Stimulus-Responsive Nanopatterned Polymer Brushes by Scanning-Probe Lithography. *Nano Letters*, Vol.4, No.2 (January 2004), pp. 373–376, ISSN 1530-6984.

Blackledge, C.; Egebretson, D. A. & McDonald, J. D. (2000). Nanoscale Site-Selective Catalysis of Surface Assemblies by Palladium-Coated Atomic Force Microscopy Tips: Chemical Lithography without Electrical Current. *Langmuir*, Vol.16, No.22 (September 2000), pp. 8317–8323, ISSN 0743-7463.

Tello, M.; García, F. & García, R. (2002). Linewidth Determination in Local Oxidation Nanolithography of Silicon Surfaces. *Journal of Applied Physics*, Vol.92, No.7 (September 2002), pp. 4075-4080, ISSN 1089-7550.

Liu, S.; Maoz, R.; Schmid, G. & Sagiv, J. (2002). Template Guided Self-Assembly of [Au$_{55}$] Clusters on Nanolithographically Defined Monolayer Patterns. *Nano Letters*, Vol.2, No.10 (August 2002), pp. 1055-1060, ISSN 1530-6984.

Jang, C. H.; Stevens, B. D.; Carlier, P. R.; Calter, M. A. & Ducker, W. A. (2002). Immobilized Enzymes as Catalytically-Active Tools for Nanofabrication. *Jounal of the American Chemical Society*, Vol.124, No.41 (September 2002), pp. 12114-12115, ISSN 0002-7863.

Kaholek, M.; Lee, W. K.; Ahn, S. J.; Ma, H.; Caster, K. C.; LaMattina, B. & Zauscher, S. (2004). Stimulus-Responsive Poly(N-isopropylacrylamide) Brushes and Nanopatterns Prepared by Surface-Initiated Polymerization. *Chemical of Materials*, Vol.16, No.19 (August 2004), pp. 3688-3696, ISSN 1520-5002.

Sugimura, H. & Nakagiri, N. (1995). Degradation of a Trimethylsilyl Monolayer on Silicon Substrates Induced by Scanning Probe Anodization. *Langmuir*, Vol.11, No.10 (October 1995), pp. 3623-3625, ISSN 0743-7463.

Amro, N. A.; Xu, S. & Liu, G. Y. (2000). Patterning Surfaces Using Tip-Directed Displacement and Self-Assembly. *Langmuir*, Vol.16, No.7 (March 2000), pp. 3006-3009, ISSN 0743-7463.

Xu, S.; Miller, S.; Laibinis, P. E. & Liu, G. Y. (1999). Fabrication of Nanometer Scale Patterns within Self-Assembled Monolayers by Nanografting. *Langmuir*, Vol.15, No.21 (August 1999), pp. 7244-7251, ISSN 0743-7463.

Liu, G. Y.; Xu, S. & Qian, Y. (2000). Nanofabrication of Self-Assembled Monolayers Using Scanning Probe Lithography. *Accounts of Chemical Research*, Vol.33, No.7 (March 2000), pp. 457-466, ISSN 1520-4898.

Pena, D. J.; Raphael, M. P. & Byers, J. M. (2003). "Dip-Pen" Nanolithography in Registry with Photolithography for Biosensor Development. *Langmuir*, Vol.19, No.21 (September 2003), pp. 9028-9032, ISSN 0743-7463.

Schwartz, P. V. (2002). Surface Functionalization and Stabilization of Mesoporous Silica Spheres by Silanization and Their Adsorption Characteristics. *Langmuir*, Vol.18, No.10 (April 2002), pp. 4014-4019, ISSN 0743-7463.

Maynor, B. W.; Li, J.; Lu, C. & Liu, J. (2004). Site-Specific Fabrication of Nanoscale Heterostructures: Local Chemical Modification of GaN Nanowires Using Electrochemical Dip-Pen Nanolithography. *Jounal of the American Chemical Society*, Vol.126, No.20 (May 2004), pp. 6409-6413, ISSN 0002-7863.

Lee, SH.; Saito, N. & Takai, O. (2009). Highly reproducible technique for three-dimensional nanostructure fabrication via anodization scanning probe lithography. *Applied Surface Science*, Vol.255, No.16 (May 2009), PP. 7302-7306, ISSN 0169-4332.

Lee, SH.; Ishizaki, T.; Saito, N. & Takai, O. (2007). Electrochemical soft lithography of an 1,7-octadiene monolayer covalently linked to hydrogen-terminated silicon using scanning probe Microscopy. *Surface Science*, Vol.601, No.18 (September 2007), pp. 4206-4211, ISSN 0039-6028.

Lee, SH.; Ishizaki, T.; Saito, N. & Takai, O. (2007). Local generation of carboxyl groups on an organic monolayer through chemical conversion using scanning probe anodization.

Materials Science and Engineering C, Vol.27, No.5-8 (September 2007), pp. 1241-1246, ISSN 0928-4931.

Saito, N.; Lee, SH.; Ishizaki, T.; Hieda, J.; Sugimura, H. & Takai, O. (2005). Surface-Potential Reversibility of an Amino-Terminated Self-Assembled Monolayer Based on Nanoprobe Chemistry. *The Journal of Physical Chemistry B,* Vol.109, No.23 (May 2005), pp. 11602-11605, ISSN 1089-5647.

Sugimura, H.; Saito, N.; Lee, SH. & Takai, O. (2004). Reversible nanochemical conversion. *Journal of Vacuum Science & Technology B*, Vol.22, No.6 (November 2004), pp. L44-46, ISSN 1520-8567.

7

Controlled Fabrication of Noble Metal Nanomaterials via Nanosphere Lithography and Their Optical Properties

Yujun Song
Key Laboratory for Aerospace Materials and Performance (Ministry of Education),
School of Materials Science and Engineering, Beihang University, Beijing,
China

1. Introduction

Since ancient time, noble metal has been used to make ornaments, jewelry, high-value tableware, utensils, currency coins and medicines due to its brilliant metallic luster, stability in air and water and anti-bacteria and anti-fungi properties (Jain, Huang et al. 2007) (Erhardt 2003; Daniel and Astruc 2004; Brayner 2008; Maneerung, Tokura et al. 2008). In fact, noble metal is also valuable due to its unique physicochemical properties, the highest electrical and thermal conductivity, the lowest contact resistance, and the highest optical reflectivity (particularly in ultra-violet region) of all metals(Edwards and Petersen 1936 ; Hammond 2000). Its d-electron configuration endows them with active chemical properties, for example, 3 variable oxidation states for silver, the most common of which is the +1 state, as in $AgNO_3$, the +2 state as in silver(II) fluoride AgF_2, and the +3 state as in compounds such as potassium tetrafluoroargentate $K[AgF_4]$, and suitability as catalysts by losing one or two more 4d electrons (Dhar, Cao et al. 2007). Silver and gold have the stable face-centered cubic (fcc) crystal structures but readily absorbs free neutrons due to its massive nucleus, which make them good absorbers for nucleus raidation. These unique features have enabled them to be applied to diverse applications such as those mentioned above, medical and dental applications, photography, electronics, nuclear reactors, catalysts, clothing and foods (http://en.wikipedia.org/wiki/Silver).

The intrinsic features of noble metal also endow their nanoscale species with attractive physicochemical properties due to the size and shape effects, including unique optical properties (e.g. Localized Surface Plasmon Resonance: LSPR; Surface Enhanced Raman Scattering: SERS), catalytic/electric properties and bio-functions (Percival, Bowler et al. 2005; Jain, Huang et al. 2007; Schwartzberg and Zhang 2008; Zhou, Qian et al. 2008; Vo-Dinh, Wang et al. 2009). Although ancient people used some features of Ag or Au nanocolloids (e.g. optical property) in fabrication of ceramic glazes for lustrous or iridescent effect in ancient Persia, they did not realize that these effects were due to nanoscale effects from size, shape and surface morphology dependent physicochemical properties of silver materials (Erhardt 2003; Brayner 2008). As materials science has progressed down to nanoscale, the unique properties of nanoscaled noble metal materials are only now being recognized and realized intentionally. These properties have shown vast applications in microelectronics, photonic devices, optoelectric coupling, catalytic processes, biomedical engineering and medicines. As understanding that the intrinsic properties (e.g. optical, catalytic) of nanomaterials on the size, shape, surface spatial morphology and arrangement (Ahmadi, Wang et al. 1996; Jensen, Duval Malinsky et al. 2000; Mock, Barbic et al. 2002; Haynes, Mcfarland et al. 2003; Noguez 2007; Song 2009) has increased, fabrication of silver or gold nanoparticles (NPs) and their arrays with controlled three-dimensional (3D) morphologies, interspacing and orientation has become a very significant research stream in recent years. A variety of fabrication techniques, such as thermal decomposition, metal salt reduction, photo reduction/conversion, template assisted growth and deposition, γ-ray-irridation, as well as microfluidic processes have been developed. As a result, significant progresses have been achieved in the tailoring of the 3-dimension (3D) morphologies (size, shape and surface morphology), crystal structures and spatial arrangement of noble metal nanomaterials as desired.

Template assisted (TA) lithography (LIGA) has developed to a powerful physical technique that enables the production of surface morphology confined NPs and NPs arrays with controlled shapes, sizes and interparticle spacing (Jensen, Duval Malinsky et al. 2000; Lee, Morrill et al. 2006; Zhang, Whitney et al. 2006; Lombardi, Cavallotti et al. 2007; Zhu, Li et al. 2008; Song 2009). Lots of templates have been developed for these purposes, such as porous polymers(Lombardi, Cavallotti et al. 2007), porous Al_2O_3 foils(Chong, Zheng et al. 2006; Lee, Morrill et al. 2006; Xu, Meng et al. 2009), or nanosphere arrays (polymers or ceramics)(Zhang, Whitney et al. 2006; Song 2009; Song and Elsayed-Ali 2010; Song , Zhang et al. 2011), resulting in varieties of template-assisted lithography, correspondingly as porous polymers LIGA (PP-LIGA), porous anodic Al_2O_3 LIGA (PAA-LIGA), or nanosphere-LIGA (NSL). Among them, the most popular and well-developed method may be NSL. In this chapter, recent progresses in NSL, for controlled producing noble metal nanomaterials will be summarized. The first discussion involves in this technique for size, shape and surface morphology controlled fabrication of noble metal nanoparticles (NPs) and nanoarrays. Then four distinct progresses in the development of NSL techniques: (1) Fabrication of hierarchically ordered nanowire arrays on substrates by combination of NSL and Porous anodic alumina (PAA); (2) Identification of single nanoparticles and nano-arrays by combination of NSL and multi-hierarchy arrayed micro windows; (3) Fabrication of biosensing system based on the combination of the noble metal nanoparticles and nanoarrays fabricated by NSL and microfluidic techniques; (4) Synthesis of solution-phased nanoparticles by the transfer of the surface confined NPs fabricated by NSL into solutions, will be discussed. In (2) and (3), the related 3D morphologies and arrangement dependent optical properties, and comparison between the numerical and experimental results, revealing their intrinsic quantum mechanism, such as LSPR will be analyzed. These researches are fundamental requirements for the discovery of novel properties and applications of noble metal NPs, as well as for paving the theory development. Finally, issues and perspectives in the controlled fabrication of noble metal nanomaterials by NSL, and investigation of their 3D morphologies and arrangement dependent optical properties for future potential applications will be highlighted and discussed in closing.

2. Size and shape controlled fabrication of nanomateials via NSL

Nanospheres have been used to form uniformly arranged layers as templates to produce perfect triangle nanoprisms on substrates (Haynes and van Duyne 2001; Song and Elsayed-

Ali 2010). The routine procedure for the production of triangular shaped nanoprisms, based on the nanosphere LIGA, is described in Figure 1 (a: cross-section view; b: top view) (Hulteen, Treichel et al. 1999; Haynes and van Duyne 2001; Song 2009). The hexagonal arranged nanosphere mono layer is first formed on the substrate by a coating process (e.g. dip-coating, rotating-coating or spinning-coating) (Step 1: a). The interstitials among any three adjacent nanospheres will form triangle shaped voids (Step 1: b) as templates. The desired noble metal (e.g. Ag) will then be deposited on the triangle shaped interstitials among the nanospheres to form triangle shaped Ag NPs (Step 2: a and b). After nanospheres are released by sonication or other methods, surface-confined triangular Ag nanoprisms can be obtained (Step 3: a). By this nanosphere LIGA process, uniform hexagonal-arrayed triangle nanoprisms can be fabricated on a variety of substrates (e.g. glass, mica, silica wafer, PMMA, etc.). Step 3-b is an Atomic Force Microscope (AFM) image of Ag triangular nanoprisms fabricated by our group using a self-assembled monolayer of 300 nm polystyrene nanospheres as the template (Song 2009).

The initial critical step in NSL is the formation of a uniform large scale nanosphere template. Both drop-coating or spin-coating can produce uniform templates on a glass, silica wafer or mica substrate. The uniformity of the nanosphere template produced by drop coating depends on the nanosphere type and concentration, the hydrophilic properties of the substrate, the environmental humidity and temperature, and the drying speed. A monolayer colloidal polystyrene nanosphere mask can be prepared by drop-coating of ~3.0-4.0 μL, 3-10 times diluted nanosphere solution (conc. 4.0 wt.%) onto the glass support and leaving them to dry overnight. A detailed procedure to fabricate the nanosphere mask using drop-coating is as follows. The glass substrates are cleaned by sonication with a mixture of sulfuric acid and hydrogen peroxide (3:1 = conc. H_2SO_4: 30% H_2O_2, Volume ratio) at 80 °C for 30 min and washed using sufficient nanopure water. Then, the glass substrates are sonicated in a mixture of ammonia and hydrogen peroxide (5:1:1 = H_2O: NH_4OH (37%): 30% H_2O_2, volume ratio) to increase the hydrophilic property on the surface of the glass substrates. Finally, the glass substrates are washed using sufficient nanopure water again and stored in the nanopure water for future use. When drop coating is to be performed, the glass substrate is picked up from the nanopure water from one of its edges. The remaining water droplets on the glass substrate are removed by touching the opposite edge on filter paper. The substrate is then left flat in a clean Petri-dish with a tilt angle of ~3-5°. A 15 μL of PS nanosphere solution is added on the surface of the glass substrate using a droplet. The water spreads over the whole glass substrate to form a semi-ellipsoidal shaped water spot. The Petri-dish is left for enough time to allow the water to evaporate. During evaporation, the temperature is kept at 18±3 °C and the humidity is kept ~50±5%. In our group, a near-uniform monolayer nanosphere template can be prepared on almost the whole glass substrate (18 mm diameter). Figure 2 shows one typical area of a near-uniform monolayer template over scale ~20 μm. From the magnified image, a selected area shown in the inset, no lattice defects can be observed. Using this template, uniform Ag nanoprisms can be fabricated by vapor deposition process. One typical area fabricated by my group is shown in Step 3-b in Figure 1, where these nanoprisms have very uniform edge length of 67 ± 4 nm (STDEV% of 6%) and thickness of 20.0±1.0 nm (STDEV% of 5%) (Song 2009).

Recent progress in nanosphere lithography (NSL) has shown that it provides a good template for other shape (besides triangle) controlled fabrication of surface confined NPs by a combination of deposition angle tilting, multi-step deposition and different post treatment methods(Haynes and van Duyne 2001; Song and Elsayed-Ali 2010). A new class of NSL structures has been fabricated by varying the deposition angle, θ_{dep}, between the nanosphere mask and the beam of material being deposited, which is hereafter referred to as angle-resolved NSL (AR NSL)(Haynes and van Duyne 2001). The size and shape of the three-fold interstices of the nanosphere mask change relative to the deposition source as a function of

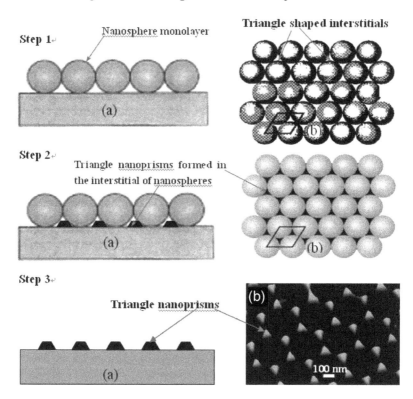

Fig. 1. The NSL process for triangular NPs fabrication. Step 1a: The hexagonal arranged nanosphere mono layer is first formed on the substrate by coating process; Step 1b: The interstitials among any three adjacent nanospheres will form triangle shaped voids as templates; Step 2a-b: the Ag metal will be deposited on the triangle shaped interstitials among the nanospheres to form triangle shaped Ag NPs; Step 3a: The nanospheres will be released by sonication or other methods, leaving the triangle shaped Ag nanoprisms on the substrates, by this nanosphere LIGA process, the hexagonal arrayed uniform triangle nanoprisms can be fabricated on variety of substrates (e.g. glass, mica, silica wafer, PMMA, etc.); Step 3b: The AFM image for Ag triangle nanoprisms fabricated by monolayer template from 290 nm polystyrene nanospheres in my group, these nanoprisms have very uniform edge length of 67 ± 4 nm (STDEV% of 6%) and thickness of 20.0 ± 1.0 nm (STDEV% of 5%). (a): cross-section view; (b) top 3D view. (Adapted in part from Song, Y. China Patent, CN200910085973.9; Haynes, C. L.; van Duyne, R. P., *J. Phys. Chem. B* 2001 105, 5599, Figure 2, Copyright (2001) American Chemical Society; and Hulteen, J. C.; et al., *J. Phys. Chem. B* **1999** 103, 3854, Figure 1, Copyright (1999) American Chemical Society.)

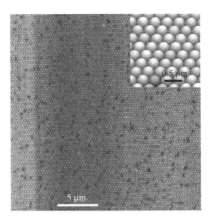

Fig. 2. Nanosphere templates based on 290 nm spherical polystyrene nanospheres for Ag nanoparticle fabrication (Reprinted from Song et al., Appl. Surf. Sci. 2010 256, (20), 5961, Figure 1. Copyright (2010) Elsevier.).

θ_{dep}, and accordingly, the deposited nanoparticles' shape and size are controlled directly by θ_{dep} and the diameter of nanosphere. Figure 3 schematically describes the effect of angle-resolved deposition on the interstices of a NSL mask from the top view. (Figure 3). As a convention, θ_{dep} = 0° represents a substrate mounted normal to the evaporation beam (Figure 3A), and all variations of θ_{dep} are made by mounting the substrates on machined aluminum blocks. It is clear from this illustration that an increase in θ_{dep} causes the projections of the interstices onto the substrate to decrease and shift (Figure 3B and 3C). At high values of θ_{dep} (e.g. 45°, Figure 3C), the projections of the interstices close, completely blocking the substrate to line of sight deposition.

One very important consequence of AR NSL, beyond the increased flexibility in nanostructure architecture, lies in the decrease in nanoparticle size. Before AR NSL, the only way to fabricate nanoparticles in the 1-20 nm size range with NSL required self-assembly of nanospheres with diameters on the order of 5-100 nm. Not only synthesis of uniform nanospheres at this range is usually difficult, but self-assembly of such small nanospheres into well ordered 2D arrays is extremely challenging because of problems with greater polydispersity and the surface roughness of substrates. However, with AR NSL, increasing θ_{dep} from 0° to 20° will halve the in-plane dimension of nanosphere templates, leading to the success in small nanoparticle preparation by NSL. In addition, nano-overlapped, nano-gapped and nano-chained structures can be addressed by multi-step AR NSL, which is fulfilled by depositing materials through a nanosphere mask mounted at different θ_{dep} below the overlap threshold value of θ_{dep} several times. Van Duyne et al have used two step AR NSL to fabricate over-lapped and gapped Ag nanostructures through a nanosphere mask with D = 542 nm onto mica substrates by a first deposition at θ_{dep} = 0° and a second deposition at an increased θ_{dep}. The importance in the fabrication of over-lapped nanoparticles theoretically exists in the enhanced optical properties due to their increased aspect ratio (in-plane width/out-of-plane height) nanoparticles(Kreibig and Vollmer 1995). Nano-overlapped structures can give an significantly increased sensitivity of optical response since they allow predictable aspect ratio to increase up to double of the original value (Haynes and van Duyne 2001). One of the interests for gapped nanostructures may exist in the investigation of the distance dependent LSPR coupling amomng gapped nanostructures.

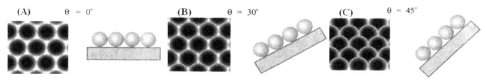

Fig. 3. Scheme of the angle resolved deposition process. (A) Samples viewed at 0°. The interstices in the nanosphere mask are equally spaced and of equal size. (B) Sample viewed at 30°. The interstices in the nanosphere mask follow a pattern including two different interparticle spacing values, and the interstitial area is smaller. (C) Sample viewed at 45°. The interstices are now closed to line of sight deposition. (Adapted from Haynes, C. L.; van Duyne, R. P., *J. Phys. Chem. B* 2001 105, 5599, Figure 5, Copyright (2001) Amercian Chemical Society.)

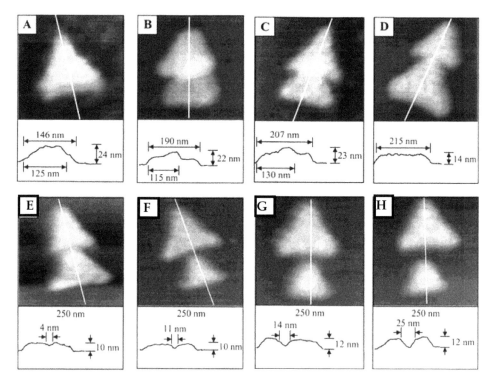

Fig. 4. Contact mode AFM images of typical Ag nano-overlapped and nano-gapped structures fabricated on mica substrates using the single layer nanosphere template. (A) 300 nm - 300 nm image, D = 542 nm, dm = 20 nm, θ_{dep} = 0° and 6°. (B) 250 nm - 250 nm image, D = 542 nm, dm = 20 nm, θ_{dep} = 0° and 10°. (C) 300 nm - 300 nm image, D = 542 nm, dm = 20 nm, θ_{dep} = 0° and 15°. (D) 250 nm-250 nm image, D = 542 nm, dm = 20 nm, θ_{dep} = 0° and 20°. (E) 250 nm - 250 nm image, D = 542 nm, dm = 20 nm, θ_{dep} = 0° and 22°. (F) 250 nm - 250 nm image, D = 542 nm, dm = 20 nm, θ_{dep} = 0° and 23°. (G) 250 nm - 250 nm image, D = 542 nm, dm = 20 nm, θ_{dep} = 0° and 24°. (H) 250 nm - 250 nm image, D = 542 nm, dm = 20 nm, θ_{dep} = 0° and 26°. (Adapted from Haynes, C. L.; van Duyne, R. P., N *J. Phys. Chem. B* 2001 105, 5599, Figure 6 and Figure 7, Copyright (2001) Amercian Chemical Society.)

The overlap percent of nanoparticles can be adjusted by θ_{dep} at a certain mass deposition thickness (e.g. 20 nm) and nanosphere diameter (e.g. 542 nm). As shown in Figure 4A-D, the overlap percent decreases with the increase of θ_{dep} from $0°$ to $20°$. The θ_{dep} at $20°$ is the threshold deposition angle since neither overlap nor gap is visible by AFM investigation at this point (Figure 4D). When the second deposition angle is more than the threshold angle (e.g. 20 ° based on the fabrication condition using nanosphere mask with D = 542 nm and mass thickness d_m = 20 nm(Haynes and van Duyne 2001)), nano-gapped structures can be formed. With the same experimental parameters defined above, the gap between nanoparticles increases as θ_{dep} is increased from $22°$ to higher values up to thecritical θ_{dep} value at which the interstitial projections are closed to line-of-sight deposition. Figure 4E-H shows the AFM images of the typical nano-gapped Ag structures with different gap distances when the second deposition angles change from $22°$, to $23°$, to $24°$ and to $26°$. Another of the applications proposed by van Duyne is to use the nanogap architecture to measure the electrical conductivity of a single molecule or nanoparticle (Haynes and van Duyne 2001). If one side of the nanogap is insulated from a conductive substrate while the other side of the nanogap is in contact with a conductive substrate, the conductance of the junction should be measurable with a scanning tunneling microscopy probe.

Clearly, like the two depositions at different values of θ_{dep}, three or more depositions will further extend the range of nanoparticle architectures accessible by AR NSL. An endless number of nanostructures are possible when one combines the ability to vary θ_{dep} and to perform multiple material depositions. As an example, the nanochain motif with three-connected-nanoparticle chains can be fabricated by three consecutive depositions. The first deposition is done at θ_{dep} = -15^0, whereas the second and third depositions will be further done at θ_{dep} = $0°$ (tilted forward) and θ_{dep} = $15°$ (tilted backward). An AFM image of the nanochain structure is shown in Figure 5 and gives a typical domain where the sample tilt axis is aligned with the triangular base of the nanoparticles. Possible applications of the nanochain architecture include sub-100 nm near-field optical waveguides, chemical and biological sensors based on the LSPR of these high aspect ratio nanoparticles, and the fabrication of nanowires.

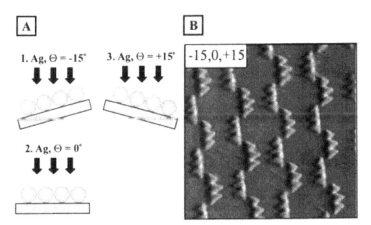

Fig. 5. (A) Schematic fabrication process and (B) contact mode AFM image for three deposition nanochain structure on mica. 1.6 μm - 1.6μm area, D = 542 nm, dm = 10 nm, θ_{dep} = +15°, 0°, and -15°. (Reprinted from Haynes, C. L.; van Duyne, R. P., *J. Phys. Chem. B* 2001 105, 5599, Figure 9, Copyright (2001) Amercian Chemical Society.)

We recently developed a modified NSL process to fabricate Ag NPs with controlled shapes on substrates. The modification in NSL is performed by thermally annealing the triangular nanoprisms, and sonication to remove weak tips, followed by removing debris and small broken parts around the NPs on the substrates(Song and Elsayed-Ali 2010). The detailed process is shown in the following: (1) Releasing the nanospheres by immersing the cover slip into a 5% HCl solution for 30 minutes, then immersing the glass substrates into CH_2Cl_2 for 30 s, then sonication for ~20-60 s; (2) The fabricated Ag nanoprisms on the glass substrates are annealed at 100-300 °C for 2-5 hours; (3) Then Ag nanoprisms are cleaned by immersing the glass cover slip into 5% HNO_3 for 10-20 s to remove any surface contamination and dissolve debris around the NPs, and then washed by large amount of nanopure water. Comparing the AFM images in Step 3-b of Figure 1 showing the NPs without above post treatment, tip-rounded triangle nanoprisms, square-shaped and trapezoidal Ag NPs (Figure 6) can be obtained via one or two of the above treatment. We observe that thermal annealing results in much more uniform NP surfaces without the thin, weak tips and edges (Figure 6(a-1)). From the magnified AFM plane image in the inset of Fig. 6(a-1) and the 3D image in Figure 6(a-2), the NPs still show triangular prism shape with

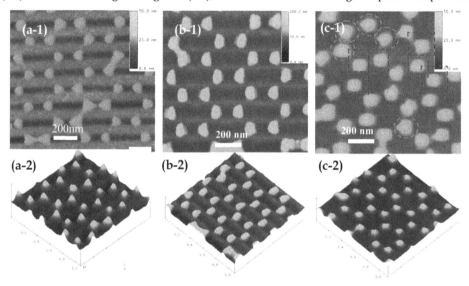

Fig. 6. Surface-confined Ag NPs with controlled shapes fabricated by the modified NSL process. (a-1) AFM image of triangular prism Ag NPs with rounded tips after thermal annealing at 200 °C for 4 hours, cleaning by 5% nitric acid, and washing by nanopure water. (a-2) The 3D image of the triangular prism Ag NPs with rounded tips. (b-1) Flat trapezoidal Ag NPs after sonication to remove one tip, thermal annealing, cleaning by 5% nitric acid, and washing by nanopure water. (b-2) The 3D image of the trapezoidal Ag NPs with one snipped tip. (c-1) The quadrilateral or pentagon shaped Ag NPs after sonication intensively to remove two tips, thermal annealing, cleaning by 5% nitric acid and washing by nanopure water. Dashed circles: pentagonal Ag NPs with one sharp tip left; dashed squares: quadrilateral Ag NPs. (c-2) is the 3D image of the quadrilateral and pentagon shaped Ag NPs. (Song Y.; Elsayed-Ali H. E., Appl. Surf. Sci. 2010 256, (20), 5961, Figure 3, Copyright (2010) Elsevier.)

rounded edges and little surface defects. Alternatively, if we sonicate the NPs produced by NSL for ~30-45 s to remove a weak tip, anneal them at 200 °C for 1- 4 hours, then wash them with 5% nitric acid, trapezoidal shaped NPs with rounded edges are formed, as shown in Figure 6(b-1) and 6(b-2). If the sonication time is increased to more than 2 min, the NPs lose their two sharp tips and form quadrilateral or pentagon shaped NPs. After thermal annealing for 1-4 hours and washing with 5% nitric acid, their edges and corners become rounded, as shown in Fig. 6(c-1), which show quadrilateral NPs (in dashed squares) or pentagon (in dashed circles). The 3D AFM image, Figure 6(c-2), shows that these NPs have rounded edges and corners. Clearly, even after thermal annealing, they are still showing prism shapes with increased thickness from their edges to centers according to their 3D AFM images.

The work described above demonstrates that NSL, broadly defined to include AR NSL and some modified post treatment after deposition of the desired materials, is manifestly capable of creating far more than arrays of nanotriangles, nanodots as previously supposed. The progresses in NSL endow much potential in the size and shape controlled fabrication of nanoparticles and nanoarrays, which gives NSL a bright future since the ability of NSL to synthesize monodisperse, size- and shape- tunable nanoparticles can be exploited to precisely investigate the size- and shape- dependent physiochemical properties of nano-optics and nanoarrays.

3. Investigation of optical properties of specific noble metal nanoparticles and nanoarrays by the combination of NSL and multi-hierarchy arrayed micro windows

The physicochemical properties of nanomaterials significantly depend on their three-dimensional (3D) morphologies (sizes, shapes and surface topography), their surrounding media, and their spatial arrangement. Systematically and precisely correlating these parameters with the related physicochemical properties of specific single nanoparticles (NPs) or nanoarrays is a fundamental requirement for the discovery of their novel properties and applications, as well as for advancing the fundamental and practical knowledge required for the design and fabrication of new materials(Song , Zhang et al. 2011). The lack of effective means of fabricating recognizable 3D morphologies controlled NPs and nanoarrays and correlating their structure parameters with their physicochemical properties as observed by different characterization techniques represents an obstacle for studying the 3D morphology-dependent properties of individual NPs and nanoarrays(Song, Zhang et al. 2011). Most current studies investigate the physicochemical properties of the NP ensemble, but not of a single NP(Jin, Cao et al. 2001; Kelly, Coronado et al. 2003; Haes, Zou et al. 2004; Song , Zhang et al. 2011). The ensemble of NPs is typically heterogeneous, because the morphologies of individual NPs prepared by routine chemical synthesis or physical vapor fabrication methods are rarely identical at the nanometre or sub-nanometre scale (Song , Zhang et al. 2011). Effective methods for 3D morphology controlled fabrication of nanomaterials, and to correlate their 3D morphology of single NPs or nanoarrays with their physicochemical properties are also essential to address fundamental and practical questions related to the single NPs (Song , Zhang et al. 2011).

An important research area in nanoscale plasmonic optics is single NP identification and characterization of their 3D morphologies and space-orientation dependent physicochemical properties(Yang, Matsubara et al. 2007; Song 2009; Song , Zhang et al. 2011). Recently, much attention has been given to the localized surface plasmon resonance (LSPR) of metal NPs because of their promising applications in plasmonic circuits, optoelectronic transducers, optical bioprobes, and surface plasmon resonance interference lithography(Shen, Friend et al. 2000; Prasad 2004; Ozbay 2006; Song 2009; Song , Henry et al. 2009; Song , Jin et al. 2010; Song, Sun et al. 2010; Song , Zhang et al. 2011). Since the plasmonic properties of metal NPs intrinsically rely on their size, shape, surface topography, crystal structure, inter-particle spacing and the dielectric environment around them, methods to correlate their plasmonic properties with the above structural and environmental parameters have become one of the most rapidly developing research directions (Song , Zhang et al. 2011).

In the precise investigation of the relationship between the LSPR properties and their 3D morphologies of specific nanoparticles and nanoarrays, two kinds of methods have been developed recently, or the *in situ* method and the spatial-localization method (Song , Zhang et al. 2011). The *in situ* method combines at least two different instruments together to conduct the structure and property characterization simultaneously: one can be used to characterize the 3D morphology (e.g. AFM or STEM) of NPs and the others will be used to chatacteize the LSPR-related optical properties of the same NPs (e.g. Dark-field microscope and spectroscopy). The spatial-localization method requires using markers to recoganize the same single nanoparticle in different instruments. We have also developed one spatial-localization method to precisely investigate the 3D morphologies dependent LSPR properties of specific NPs and nanoarrays by the combination of NSL and traditional UV-LIGA, where Ag NPs and nanoarrays can be fabricated by NSL in the pre-formed multi-hierarchy arrayed transparent micro-windows on the substrates (e.g., glass cover slip) by the UV-LIGA(Song 2009; Song , Zhang et al. 2011). This technique permits easy characterization of the 3D morphologies of single NPs by AFM or SEM and their LSPR spectra using dark-field optical microscopy and spectroscopy (DFOMS). It is also possible to investigate the local morphology dependence of the LSPR spectra of the single NPs and nanoarrays. In this method, multi-hierarchy arrayed micro windows are first fabricated on a glass cover slip using the standard photolithography, whose details are shown in reference 27. Fig. 7A and Fig. 7B show one example of the designed multi-hierarchy arrayed micro windows (3 tiers) and the typical final micro-windows (Fig. 7C) pattern after printing. The multi-hierarchy arrayed micro-windows on the glass cover slip are used to identify the location and orientation of single NPs, whose tiers can be determined by the observed field at desired resolution. For example, in the first tier of the multi-hierarchy arrayed micro windows (Fig. 7A), each local area can be discerned by marking its X and Y number, such as the shaded area X1–Y2. Then, in the second tier of the multi-hierarchy arrayed micro windows (Fig. 7B), the scale can be reduced by M or N times and each local area can also be marked by x and y number. If this area is the sub-tier in the shaded area of the first tier, it can be labeled as X1–Y2–x3–y3. In a similar way, step-by-step, we can reach the last tier with several transparent micro windows available (Fig. 7C), in which the desired nanoparticle can be made by different fabrication methods (e.g., electron beam lithography or nanosphere lithography). Nanoparticles less than 10 nm of different shapes synthesized by a wet-chemical process can be immobilized by a routine diluted deposition process. Consequently, the same nanoparticle in each window can be identified by comparing the images taken by the optical microscope with those characterized by the AFM. Finally, in each window, the same nanoparticle can be characterized by different techniques (e.g., DFOMS and AFM) allowing correlation of its 3D morphology with its optical response(Song 2009; Song , Zhang et al. 2011).

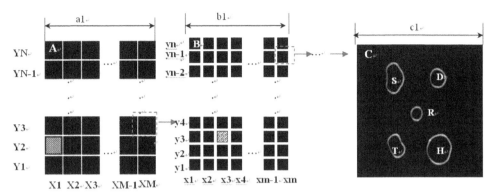

Fig. 7. The multi-hierarchy arrayed micro windows on the substrate (e.g., glass cover slip). (a) The first tier of the multi-hierarchy arrayed micro window, each local area can be discerned by marking its X and Y number, such as the red-dashed square area of X1–Y2. (b) The second tier of the multi-hierarchy arrayed micro windows, whose scale can be reduced by M or N times, whose local area can also be marked by x and y numbers. If this net area is the sub-tier n the red area of the first tier, it can be labeled as X1–Y2–x3–y3. Step-by-step, the last tier with several unique-shaped transparent windows can be reached. The open windows can be made with different shapes. (c) The nanoparticles can be fabricated on the micro-pattern by various methods (e.g., nanosphere lithography). In each window, the same nanoparticle can be identified by comparing the images taken by optical microscopy, AFM, or other microscopy methods. Finally, the structural parameters (size, shape, orientation, interparticle spacing, and thickness) can be correlated with their optical responses (Reprinted from Song Y.; et al., Nanoscale 2011, 3, 31-44, Figure 7, copyright (2011) from the Royal Society of Chemistry.)

A typical example to identify NPs and nanoarrays using both AFM and DFOMS is illustrated in Figure 8. Triangular Ag NPs and hexagon-arranged nanoarrays fabricated on the surface of glass cover slips within the nearly circle-shaped micro window can be identified and characterized using AFM (Figure 8A, 8B is the 3D AFM image of the dash-squared area in 8A) and DFOMS equipped with a color camera (Figure 8 C) and charge-coupled device (CCD) camera (Figure 8D). The CCD camera offers higher spatial resolution than the color camera, while the color camera provides the real colors of individual Ag NPs that are generated by LSPR. The center of each individual NP in the optical images recorded by the CCD is located with a single-pixel resolution (each pixel can be 125 nm or 67 nm depending on the CCD resolution and equipment setup) by determining the address of the pixel with the highest intensity of the NP. The positions of individual NPs of interest (e.g. the circled one) within the micro window in the optical images (Figure 8C and D) are then determined with a spatial resolution limited by the optical diffraction limit (~200 nm) and an orientation angle resolution of about 1.0 degree. This approach allows us to correlate AFM images of individual NPs (as the one circled in each image) with the same NP shown in its corresponding optical image and to investigate its 3D morphological-dependent LSPR properties. Clearly, these triangle nanoparticles in this window almost show the same scattering color (Figure 8C) and intensity contrast (Figure 8D). By comparing their scattering color images (Figure 8C) with their AFM images (Figure 8A and B) of these nanoparticles, it is once again showing that NSL is powerful method in the fabrication of uniform triangular nanoparticles and nanoarrays.

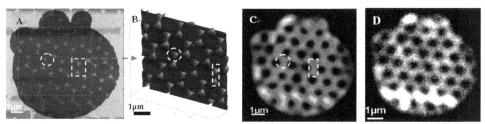

Fig. 8. One example for the identification of the specific nanoparticles and nanoarrays in different instruments via multi-hierarchy arrayed micro windows based on Ag triangle nanoparticles and nanoarrays fabricated by NSL in one nearly circle-shaped window. (A) The plan view of the hexagon-arrayed triangle Ag NPs in one circle micro-window scanned by atomic force microscope (AFM); (B) the 3D view of the hexagon-arrayed triangle Ag NPs marked in the large pink dash-square in (A); (C) the real scattering color of these hexagon-arrayed triangle Ag NPs observed under a dark-field microscope; (D) the CCD images of the scattering light of these hexagon-arrayed triangle Ag NPs recorded by a CCD camera equipped in the dark-field microscope. The dashed circles in each image refer to the same specific particle and the dashed squares in each image refer to the same specific nanoparticle pair. (Adapted from reference Y. Song, China Patent, Appl. No. CN200910085973.9).

We have used it to investigate size- and shape-dependent LSPR spectra of single Ag NPs by the analysis of the experimental results with the theoretical calculation (i.e. DDA simulation)(Song , Zhang et al. 2011). Figure 9 gives the AFM images of one specific triangle-shaped Ag NPs characterized by multi-hierarchy arrayed micro windows. The AFM image of the triangular silver NP shows that it has the edge length of 375-420 nm (Figure 9A) and the out-of-plane height of about 16.1 nm (Figure 9B). This NP shows multi LSPR scattering colors (Figure 9C), as further evidenced by its multi-mode LSPR peaks at 562.3 nm, 659.9 nm and 759.6 nm (Figure 9D-b). The peak wavelengths, peak ratios, and line widths (FWHM) at 562.3 nm and 659.9 nm from experiment are in good agreement with DDA simulation for its LSPR scattering (Figure9D-c), as have been summarized together with other shaped nanoparticles fabricated by the modified NSL in reference 27. In general, the DDA simulation shows best agreement with the experimental spectra for NPs, hence their shapes can be accurately modeled. However, it can also be seen that for wavelengths longer than 650 nm for the investigated NPs, the experimental result has a lower intensity than the simulation(Song , Zhang et al. 2011). By analysis the instrument errors and the wavelength dependent CCD quantum efficiency, these deviations are deduced by the precision in the shape construction during DDA simulations. From these results, it was also found that when the shapes and 3D morphologies of the NPs became more complicated, the deviation between the DDA simulation and the experimental result increased (Song , Zhang et al. 2011). This is due to the geometrical deviation between the real NPs and the regular species used in the calculations. If these two instrumental factors and the geometrical deviation of NPs are considered, the corrected experimental results will match with the DDA simulation very well. This result also confirms that our experimental method (DFOMS), based on the far field detection, preserves the ability to detect the near-field LSPR signal.

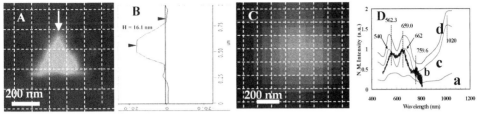

Fig. 9. (A): The AFM image for one single triangular shaped Ag nanoparticle with edge length of 375-420 nm; (B) the height mapping of the triangle shape Ag nanoparticle along the direction of the arrow in Fig. 9(A), showing the out-of-plane height of this nanoparticle of 16.1 nm; (C) the real scattering color image of this triangular shaped Ag nanoparticle; (D-a): the LSPR absorption spectrum of this nanoparticle by DDA; (D-b) the LSPR scattering by experiment; (D-c) the LSPR scattering by DDA; (D-d): the LSPR extinction by DDA. In order to identify the location and the orientation of these positions around the NPs, the AFM image and color image were netted by dashed lines with each square unit of 125 nm ×125 nm after their distances and orientations were corrected. (Adapted from references: Y. Song, China Patent, Appl. No. CN200910085973.9; Song Y.; et al., Nanoscale 2011, 3, 31-44, Figure 13, copyright (2011) from the Royal Society of Chemistry. Adapted with permission.).

This combined method based on the NSL and the multi-hierarchy arrayed micro windows also allows us to investigate the 3D morphology dependent tip-tip LSPR coupling of triangular nanoparticle pairs. The zoom-in AFM image for the detailed 3D morphology of one typical Ag nanoprism pair is shown in Figure 10A. The nanoprisms have almost the same edge size ~ 375 nm and maximum out-of-plane height ~ 17.1 nm shown in Figure 10B by the typical height map along the arrowed tip-tip direction in Figure 10A. The real scattered color for the nanoprism pair, taken from dark-field microscopy, is shown in Figure 10C. Both of the nanoprisms in the pair give red color with different brightness, which might be due to variation in their surface roughness, slightly difference in the underlying surrounding dielectrics, and the focusing distance during image recording. The middle area between the two nanoprisms clearly shows more reddish color than the optical centers of the two nanoprisms. The LSPR spectrum for the middle area of the two optical centers (representing the tip-tip-coupling) is recorded in Figure 10D using their CCD image (not shown here) for the location identification, together with that obtained by the discrete dipole approximation (DDA) calculation of the nanoprism pair. According to its 3D morphology of the nanoparticle pair, the two nanoprisms can be treated as regular triangular nanoprisms with the bottom edge length of 375 nm, the top edge length of 125 nm and out-of-plane height of 17.1 nm for conducting the DDA calculation of the nanoprism pair. The recorded LSPR spectrum (Figure 10D: a) at the middle optical center of the two nanoprisms shows three distinct peaks, one strongest peak at 605 nm, one shoulder at 536 nm, and one secondary strong peak at 754 nm. By comparing the experimental result for the tip-tip coupling of the nanoprism pair with the DDA calculation (Figure 10D: b), it can be deduced that the peak at 605 nm represents the in-plane quadrupole resonances originated from the two source nanoprisms and the peak at 536 nm is from the out-of-plane quadrapole resonances of the two source nanoprisms. Although the DDA simulation does not show one distinct peak at 754 nm, our experiment result suggest one strong peak at this wavelength, which is probably from the strong tip-tip coupling. In order to reveal whether the peak at 754 nm is mainly from the tip-tip coupling or not, the LSPR spectra from the optical centers of the source nanoprisms, are recorded (not shown here), showing one strong peak at the same position. Generally, one can see that the peak positions and shape resonances for the two nanoprisms are almost the same, suggesting that the nanosphere lithography process is very powerful in the fabrication of the nanoprisms with almost identical 3D morphologies and surroundings. Both of the two triangle nanoprisms do not give the peak at 754 nm as strong as the pair, confirming that the additional peak at 754 nm indeed is from the tip-tip coupling. However, previous investigations did not show additional strong peak due to tip-tip coupling (Su, Wei et al. 2003; Zhao, Kelly et al. 2003). The reason for this significant coupling between the nanopair may be caused by the unique size of our particles that is just lying in the range of half wavelength of visible light, which can cause a strong long-range electrodynamic interaction among light and the collective electrons on the particle surfaces.

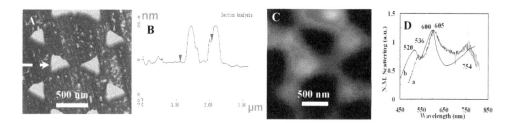

Fig. 10. Tip-tip coupling of one typical pair of triangular Ag nanoprisms in the arrays with interspacing of 103 nm is characterized using (A) AFM; (B) the scan of out-of-plane height of the two nanoprisms along the arrow direction in (A); (C) the color image taken from dark-field microscopy; (D: a) the LSPR scattering spectrum at the central locations (3 pixels) of the source nanoprisms and (D:b) the LSPR scattering spectrum of the pair calculated by DDA.

In additon, our experimental observations show that nanoprism coupling does not affect the quadrupole mode in LSPR significantly, resulting in little shifts in the highest peak at 598-605 nm (the in-plane quadrupole mode). However, one additional peak (i.e. 754 nm) as compared with the in-plane quadrupole mode can be observed. This peak resulted from LSPR coupling is in good agreement with the prediction by the semianalytical model by Schatz et al. (Zhao, Kelly et al. 2003) In the present study, the edge lengths of the triangular nanoprisms are more than $\lambda/2\pi$ (64-128 nm), which is more than the critical scale in the semianalytical model in the DDA.(Zhao, Kelly et al. 2003) Therefore, the long-range electrodynamic interaction, not electrostatic effects, will be dominant in the LSPR of the two nanoparsms. The center-to-center interspacing of the two nanoprisms is ~ 532 nm, more than the critical interspacing. As a consequence, the coupling will be mainly dertermined by the long-range radiative dipolar interactions (or radiative damping effects),(Zhao, Kelly et al. 2003) and phase retardance effects,(Su, Wei et al. 2003) resulting in one new peak with wavelength more than the highest peak for the two nanoprisms.

Based on this combined method, we have investigated the distance dependent tip-tip coupling between triangular Ag nanoprism pairs with dimensions at the range of half wavelength of visible light and distance ranging from 100 nm to 400 nm. It has been found that the coupling peak wavelength increases and the coupling intensity decreases with the increased tip-tip distance, and finally the coupling disappears (no coupling peak) when the tip-tip distance is more than about 400 nm due to the coupling intensity becomes extremely low.

Generally, the combination of NSL and the multi-hierarchy arrayed micro windows fabricated by the routine UV-LIGA shows a powerful ability not only in the identification of nanoparticles and nanoarrays but also in the precise investigation of the fundamental theory related to the 3D morphology dependent LSPR and LSPR coupling. In our study, the detector is far-field while the DDA calculation is based on the near-field. Thereby, the results indicate that the near-field LSPR of single NPs and the coupling signals of nanoarrays can be detected by the far-field detector if the 3D morphologies of NPs or nanoarrays can be precisely accounted for in the DDA model.

4. Microfluidic biosensing system based on NSL and microfluidic reactor fabrication

Recently, Song has developed a high-throughput single Ag NPs biosensing device by coupling a variety of functionalized Ag NPs fabricated by NSL into a series of microfluidic channels (Song 2009). The designed microfluidic biosensing system based on Ag single nanoparticles and nanoparticle arrays is illustrated in Figure 11. Samples were fabricated by the combination of NSL and the traditional UV-LIGA process for the microfluidic reactor fabrication (Song 2009; Song 2010; Song and Elsayed-Ali 2010). In this biosensing system, the corresponding microfluidic channels are fabricated on the designed patterns where series of single Ag NPs or arrays (Figure 11: a) have been fabricated by careful alignment. The glass cover is then connected with glass optical fiber binding on the top of the microfluidic channels after careful alignment with the desired single Ag NPs or nanoarrays. In order to alleviate the non-specific absorption in the microfluidic channels, the channels are modified by polyvinylacholol (PVA) or polyethylene glycol (PEG) solution. After that, the single Ag NPs will be surface modified by a mixture of at least two thiol compounds with one having carboxyl group or amine group as the conjugating compound (e.g. 11-mercaptoundecanoic acid: MUA), and another thiol compound without carboxyl group or amine group as spacer (e.g. 6-mercapto-1-hexanol: 6-MCH, 1-octanethiol: 1-OT). The modification reaction is shown in equation (1). The Ag NPs can then be functionalized with biomolecules, as reporter (e.g. IgG), by a conventional 1-ethyl-3-(3- dimethylaminopropyl)-carbodiimide (EDC) coupling process to form the f, as shown in equation (2) and (3) for the functionalization of Ag NPs (Figure 11: a and b).

The number per Ag NPs can be controlled by the ratio of the conjugation compounds and spacers, which can be used to calculate the number of the responding biomolecules (e.g. Protein A) that can bind with the reporters, which can be directly sensed by the LSPR peak shift. As shown in Figure 11, the solution having a specific concentration of the corresponding detected biomolecules can be delivered into the microfluidic channels (Figure 11: g). The channel widths are designed from several hundreds micro meter to ten micrometers that will play a role like a dark-field condenser for incident white light. The scattering color changes and the LSPR spectrum variations of Ag NPs (a) caused by the binding of the detected biomolecules on the reporters, (b) will be collected in the opened windows, (d) and transported into the detector and analyzer, (f) by the glass fiber, (e) after signal magnification.

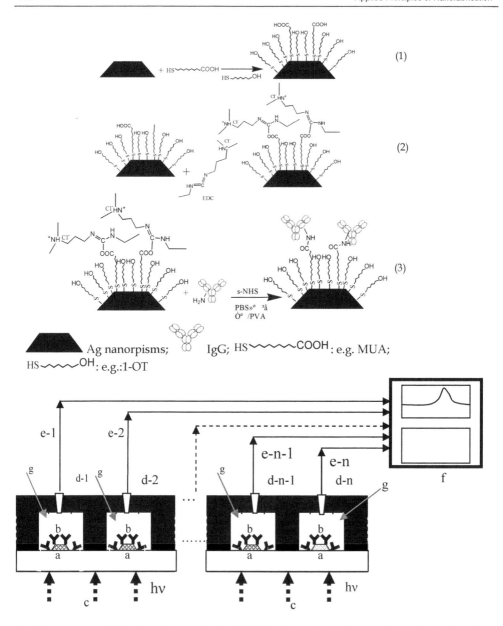

Fig. 11. a: nanoparticles (e.g. triangular Ag NPs); b: biomolecule (e.g. antibody); c: induced light (e.g. white light from tungsten lamp); d-1 - d-n) : Series of detecting windows connecting with optical microfibers in different microchannels ; e-1 – e-n : glass microfibers for signal transport ; f : Optical spectroscopy (e.g. micro optical fiber spectroscopy-S2000, Spectropro-150, surface enhanced Raman spectroscopy; g: the sealed microchannels). (Adapted from reference Y. Song, China Patent, Appl. No. CN200910085973.9)

Song has investigated the efficiency of this kind of biosensing system. Using the color change and LSPR spectra shifts based on the binding of one model biomolecule pairs (antibody: IgG is firstly functionalized on the Ag surfaces by EDC process, then the antigen: protein-A buffer solution is pumped into the microfluidic channels) as model, it can be seen that the spectra shift and the color changes from the binding of the model biomolecule pairs depend on the concentration of biomolecules and the running time. Up to now, the detection resolution of this kind of biosensors based on the scattering from single Ag NPs has reached 2 nm peak shift per 1 nM concentration change and the resolution for one single NP biosensor is calculated as 10-20 biomolecules per Ag NP(Song 2009). This result suggests another persepctive application by the combination of NSL, microfluidics and bio-functionalization process.

5. Fabrication of hierarchically ordered nanowire arrays on substrates by combination of NSL and Porous anodic alumina (PAA)

In some applications of nanomaterials, the NPs need to be arranged in some particular patterns, architectures or motifs with controlled interspacing, or conjugated with some other kinds of materials (e.g. polymers) (Chong, Zheng et al. 2006; Song , Zhang et al. 2011). The controlled arrangement and immobilization of Ag NPs on substrates will be very crucial to enable some fascinating and delicate applications, particularly in electronic circuit based electro-optical devices and long term functional composites for biological applications. Many methods have been explored for this purpose. Among them, template-assisted LIGA or structure controlled artificial fabrication methods (e.g. E-beam LIGA, NSL, PAA-LIGA) may be the most convenient techniques(Song , Zhang et al. 2011). In the NSL development, the suitability and powerful ability in the architecture and interspacing controlled fabrication of NPs and nanoarrays can be expanded extremely if the NSL can be combined with other template-assisted LIGA methods. Here we just show one example to fabricate hierarchically ordered nanowire arrays on substrates by the combination of NSL and porous anodic alumina (PAA) LIGA(Chong, Zheng et al. 2006).

Like NSL, porous anodic alumina (PAA) templates have attracted intense attention in nanodevice-oriented fabrication in recent years(Xu, Meng et al. 2009). As a well-developed template, PAA offers amazing simplicity and convenience for nanofabrication due to the capabilities of forming high-density, well-aligned, and hexagonally packed sub-100-nm pores, the ability to control the 3D pore structures by simply varying the anodization conditions, and the ease of selectively removing the template after fabrication(Chong, Zheng et al. 2006). As shown in Figure 12, one typical PAA-LIGA process includes(Lombardi, Cavallotti et al. 2007): (a) formation of a 300 nm thick PAA film on Al by a two step anodization process in 0.3M oxalic acid; (b) dissolution of unoxidized Al; (c) barrier layer etching in 5wt% phosphoric acid; (d) transfer of the PAA mask onto Au-coated Si followed by a thermal treatment to improve the adhesion of the films to the substrate; (e) Ag electrodeposition through the PAA pores; (f) PAA mask removal. By carefully controling the sizes and interpore spacing of the nanoholes, very uniform Ag nanorods with controlled interspacing can be fabricated by electroplating. The typical Ag nanorods prepared by this template assisted electrodeposition process can give a much uniform size and interparticle spacing distribution, with a standard size deviation less than 5% and a spacing deviation less than 7%.

Progresses in PAA-LIGA have shown its abilities not only in the synthesis of the traditional high aspect ratio nanomaterials, such as nanorods (Pan, Zeng et al. 2000; Lombardi, Cavallotti et al. 2007; Xu, Meng et al. 2009)and nanowires (Hong, Bae et al. 2001; Xu, Zhang et al. 2005), but also some unique nanostructures, such as porous metallic nanorods by galvanic exchange reaction (Mohl, Kumar et al. 2010), Y-junction nanowires and multiply branched nanomaterials due to its flexibility in the pore structure control according to the principle $\frac{1}{\sqrt{n}} \times V_s$, where V_s is the anodizing voltage for stem pores and n is the number of branched pores from that stem(Meng, Jung et al. 2005; Xu, Meng et al. 2009).

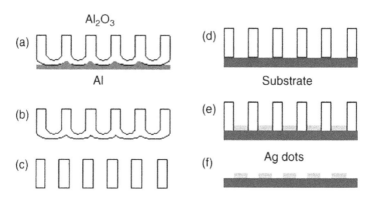

Fig. 12. Fabrication scheme for Ag nanoparticle arrays: (a) formation of a PAA film on Al by a two step anodization process in 0.3Moxalic acid; (b) dissolution of unoxidized Al; (c) barrier layer etching in 5 wt.% phosphoric acid; (d) transfer of the PAA mask onto Au-coated Si followed by a thermal treatment to improve the adhesion of the films to the substrate; (e) Ag electrodeposition through the PAA pores; (f) PAA mask removal. (Lombardi I.; et al., Sensors and Actuators B: chemical 2007, 125, 353-356, Figure 1. Copyright (2007) Elsevier.)

The combination of NSL and PAA-LIGA has been further developed to create hierarchically ordered nanowire arrays, as schematically shown in Figure 13A(Chong, Zheng et al. 2006). A monolayer of self-assembled polystyrene nano or microspheres as masks is first used to deposit periodic porous gold films on silicon substrates (Figure 13A: i-ii). Next, PAA films are fabricated on top of the porous gold film/substrate (Figure 13A: iii). Nanowires are then selectively electrodeposited into the pores of the alumina using the porous gold film as a working electrode (Figure 13A: iv-vi). In detail, a drop of polystyrene sphere suspension (e.g. 1 µm in diameter, 10 wt % aqueous dispersion) is spin-coated onto a pretreated substrate (e.g. Si, glass or mica) to form close-packed microsphere monolayers. The size of the nano or microspheres can be tuned using O_2 reactive ion etching (RIE) with a suitable O_2 flow (e.g. 20 SCCM, SCCM denotes cubic centimeter per minute at STP) at a certain pressure (e.g. 15 mTorr) and a power density (e.g. 110 W) for 6–10 min. After RIE, isolated nano or microsphere monolayers with tunable spacing will be formed. Then, about 5 nm Ti (as an adhesion layer) and 40 nm gold films in turn deposit onto the substrate using the RIE reduced nano or microspheres as masks. After removal of the mask by sonicating in a solvent (e.g. toluene) for 3 min, a porous gold film will be formed on the substrate. After that, an aluminum film with a thickness of ~500 nm deposits onto the porous gold film. Prior to anodization, the aluminum film is subjected to an imprinting step, in which a free-standing PAA with a thickness of ~10 μm fabricated by the process in Figure 12 is used as a mold for imprinting. Finally, the imprinted aluminum film is anodized in 0.3M oxalic acid at 2 °C and the barrier layer at the bottom is removed in 5 wt % H_3PO_4 for 60 min. The gold nanowires can be deposited at −1.0 V versus standard calomel electrode (SCE) from a commercial bath (Orotemp 24, Technic) for varying amounts of time. Alumina templates can be removed in 1M KOH for some time (e.g. 10 min) to obtain hierarchically patterned free-standing nanowires or nanorods.

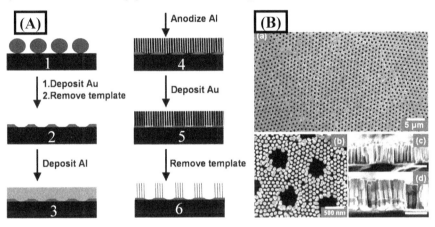

Fig. 13. (A: 1-6) Schematic of method to create hierarchical nanowire arrays on substrates. (B) SEM images of hierarchical nanowire arrays on substrate:(a) Top view of the nanowire arrays with hexagonally organized microvoids over large areas; (b) High magnification SEM image of (a); (c) Side view of the cleaved sample from (a), the concaves caused by voids are clearly apparent; (d) Side view of Au/Ni/Au/Ni segmented nanowire arrays. The side view shows clear contrast; brighter segment is gold portion. Scale bars in (c) and (d): 500 nm. (Chong, M. A. S.; et al., Appl. Phys. Lett. 2006, 89, 233104. Figure 1 and Figure 3. Copyright (2006) from American Institute of Physics. Adapted with permission.)

Figure 13B (a) shows the top view of one typical nanowire arrays with hexagonally organized microvoids (e.g. 500 nm in diameter) over large areas (Chong, Zheng et al. 2006). One of its high magnification images as shown in Figure 13B(b) clearly demonstrates the arranged patterns by individual gold nanowires. The nanowires with uniform size, replicated from the PAA, are hexagonally packed at the nanoscale. The hierarchical nanowire arrays standing on the substrate are further checked from a cleaved sample by side view SEM images (Figure 13B(c), suggesting the well aligned nanowires and the concave features from the nanoscale voids. In addition to fabricating pure gold nanowire arrays, this combinational template is also suitable for selectively electrodepositing other materials for functional device applications. In particular, the vertical structure/composition along the length of the nanowire is also tunable. For example, multilayer Au/Ni/Au/Ni nanowire arrays with in-plane hierarchical structure can be fabricated by the alternating deposition twice of nickel from a Watt's bath (300 g/L $NiSO_4$ $6H_2O$, 45 g/L each H_3BO_3 and $NiCl_2$ $6H_2O$) and of gold at −1.0 V versus SCE electrode. Figure 13B(d) shows the cross section of Au (bottom, brighter segment) /Ni (darker segment)/Au/Ni (top) nanowire arrays.

6. Solution phased nanomaterials by the releasing of nanoparticles fabricated by NSL

Recent progress in nanosphere lithography (NSL) has shown that this method provides a good template for shape-controlled fabrication of surface confined NPs(Zhang, Whitney et al. 2006; Song , Zhang et al. 2011), which also allows for flexible functionalization of these NPs on the clean surface (as cartooned in Figure 14A) using the routine functionalization process from equation 1 to 3. After the surface functionalization of surface confined nanoparticles fabricated by NSL, they can be dislodged into solution phase (as schemed in Figure 14B). This dislodging process provides a useful alternative to synthesis uniform solution phase NPs besides the wet chemical process. Van Duyne *et al.* have developed this process and used it to fabricate solution phase NPs in ethanol(Amanda, Zhao et al. 2005). However, their results indicated that most of the NPs in the solution have nonuniform surface morphologies with truncated tips in addition to the presence of debris and some of the NPs attached together on the glass substrate surface causing the agglomeration of the released NPs. In addition, aqueous phase NPs are expected to be more biocompatible than those in ethanol. Therefore, technology development to obtain aqueous-stable nanocolloids via surface modification and releasing of the surface-confined NPs fabricated by NSL into water solution are much desired.

Our group recently developed a modified NSL process to fabricate Ag NPs with controlled shapes on glass substrates and with the ability to release them into the aqueous solution without any obvious agglomeration(Song and Elsayed-Ali 2010). Three modifications of the standard procedure of nanosphere lithography were made in order to obtain stable NPs with different shapes. The modification to the process were the following: (1) Releasing the nanospheres by immersing the cover slip into a 5% HCl solution for 30 minutes, then immersing the glass substrates into CH_2Cl_2 for 30 s, then sonication for ~20-60 s; (2) The fabricated Ag nanoprisms on the glass substrates were annealed at 100-300 °C for 2-5 hours then cleaned by immersing the glass cover slip into 5% HNO_3 for 10-20 s to remove any surface contamination and dissolve debris around the NPs, and then washed by large amount of nanopure water; (3) The glass substrates were immersed into 5-10wt.% HF and HCl acid mixture (HF: HCl = 1:1) for 30-60 s or 10% NaOH solution for 60-120 s to etch part of the glass substrate under the Ag NPs, and then the substrates were washed with sufficient amounts of nanopure water. Finally, the glass substrates with the Ag NPs were dried by inert gas flow and kept in desiccators. The surfaces of the Ag NPs can be modified by chemicals containing thiol groups (such as 1-OT, MUA, 6-MCH and Tiopronin (TP)) forming strong sulfur-silver covalent bonds. We used 1-OT and MUA as functional reagents. The functional solution was prepared by dissolving 0.049 g 1-OT and 0.073 g MUA into 100 mL pure ethanol in a 100 mL volume certificated flask to form 2mM 5:1 1-OT /11-MUA solution. The Ag NPs were once again cleaned using 5 % nitric acid and then immersed into the 2 mM 5:1 1-OT /11-MUA solution and left overnight. The releasing aqueous solution contains 5 V% of 2mM 5:1 1-OT /11-MUA in nanopure water. The glass substrates with surface modified Ag NPs were removed from the functional solution and immersed into the releasing solution. The NPs was then sonicated for 30-120 s to remove them from the substrates into the releasing solution. For a 2-4 mL releasing solution, 4-8 glass substrates were used in order to reach a NP concentration suitable for optical property measurements.

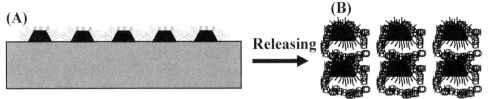

Fig. 14. (A) the surface-confined triangular Ag NPs are functionalized by chemicals with thiol groups and (B) can be further released into water or other solvents forming solution-phased nanocolloids. (Adapted from reference Y. Song, China Patent, Appl. No. CN200910085973.9)

Comparing with those Ag NPs via the traditional dislodging method(Amanda, Zhao et al. 2005; Song and Elsayed-Ali 2010), the shape integrity of the heat-treated NPs after releasing them into water can be retained perfectly. Figure 15(a) shows TEM image of the Ag NPs after thermal annealing without pre-sonication. Most of those Ag NPs show triangular shapes with rounded tips (doted circles in Fig. 15(a)) and some with snipped tips (dashed circles in Fig. 15(a)). The inset is a magnified image of these NPs, clearly showing a triangular shape with rounded tips. The histogram for these Ag NPs (Fig. 15(b)) gives a mean size of 39.6 ± 4.9 nm with much narrower size distribution of STDEV % = 12.4 % than those obtained from surface-confined Ag NPs without any post annealing (STDEV % = 41.7 %)(Song and Elsayed-Ali 2010). Fig. 15(c) is a TEM image for Ag NPs that were thermally annealed after removing two tips by sonication, whose histogram gives a mean size of about 33.9 ± 6.8 nm (Fig. 15(d)), less than that for those triangular shaped NPs with rounded tips after post-annealing. Most of these NPs show quadrilateral shapes (dashed circles) or pentagon shapes as shown more clearly in the inset of Fig. 15(c). These NPs have a similar shape as those observed by AFM images in Figs. 6(c-1) and (c-2). From the TEM images in Fig. 15(c), some of the NPs give less contrast in their central parts (NPs labeled by dashed circles). We believe that the lighter centers in these NPs are from a thinner center resulting from adhesion of the center of these NPs to the glass substrate during annealing. AFM observation of the glass substrate after removal of the NPs show debris forming hexagonal shaped arrangements. This observation is consistent with adhesion of the central part of the triangular nanoprisms to the substrate.

Variations in the shape, surface modification and surrounding environment of these NPs give significant shifts in their UV-vis absorption spectra for the surface confined NPs before and after tip rounding, after surface modification, and after dislodging into water, as shown in Figure 16. The absorption spectrum for the surface-confined Ag NPs fabricated by NSL without any tip rounding and surface modification has two distinct peaks at 476 nm and 672 nm (Figure 16a). The absorption peak at 476 nm is primarily from the higher-order mode surface plasmon resonance (e.g., quadrupole) of the NPs, and the peak at 672 nm is mainly from the dipole resonance of the NPs. We note that the higher-order resonance peak has almost the same intensity as that for the dipole resonance for all types of NPs, although the higher order modes are expected to be much weaker than the dipole resonance. Since the substrate is continuously covered by a hexagonally arranged array of Ag NPs with tip-tip distance less than 100 nm, we postulate that the particle-particle coupling will contribute to the LSPR spectrum. This particle-particle interaction effect could be responsible for the observed spectrum. When the tips in the Ag triangular nanoprisms are rounded, the tip-tip

LSPR coupling effects are alleviated, as indicated by the disappearance of the peak at 672 and the red-shift of the peak at 476 nm to 504 nm representing the higher-order surface plasmon resonance mode (Figure 16b). The absorption spectrum for the surface modified Ag NPs (Fig. 16c) shows a slight blue shift at the peak of 476 nm (to 470 nm) and a significant blue shift at 672 nm (to 626 nm) with reduced intensities. This spectrum was expected to give a red shift due to the increased dielectric constant from the adsorbed thiol compounds(Amanda, Zhao et al. 2005). We attribute this blue shift to shape variation (e.g., increased height, smooth surface topography) during surface modification by immersion that was similar to solvent annealing which results in blue-shift of LSPR since any solvent annealing has not been done on our NPs(Jensen, Duval Malinsky et al. 2000; Malinsky, Kelly et al. 2001). These variations have been observed by the slightly reduced NP size and rounded shapes observed in the TEM image of Figure 15 when compared with the AFM image of Figure 1 and Figure 6. In addition, when the Ag NPs are covered by thiol groups, the surface free electron density may be reduced, leading to weaker surface plasmon resonance in single NPs and surface plasmon resonance coupling among nanoparticle arrays(Kelly, Coronado et al. 2003). This will result in a blue shift of the LSPR peak and a reduced LSPR intensity.

The UV-vis absorption spectrum of the Ag NPs after release in water, shown in Fig. 16d, was compared to other surface confined NPs. The aqueous Ag NPs give a main peak at 532 nm and a very weak peak at 352 nm. The main peak at 532 nm appears to be from LSPR by the triangular nanoprisms with rounded tips and is blue shifted from that obtained for NPs with a LSPR peak at 605 nm fabricated by the routine NSL and released from the surface. This is attributed to the reduced size and rounded tips. The peak at 352 nm in Fig. 16d becomes much weaker and narrower than that for the aqueous Ag NPs released from the surface confined Ag NPs as fabricated by routine NSL, obviously due to the shape variation of NPs and almost no small spherical shaped debris observed in the aqueous Ag NPs released from the surface-confined Ag NPs fabricated by the modified NSL (Figure 16a). By comparing the TEM images for the two kinds of Ag NPs, it can be deduced that the peak at 352 nm in Fig. 16d is mainly from the out-of-plane quadrupole resonance of Ag nanoprisms with rounded tips according to the previous investigation(Jin, Cao et al. 2001; Amanda, Zhao et al. 2005; Zhang, Li et al. 2005). The peak intensity ratio between the main peak at 532 nm and the weak peak at 352 nm for these NPs is ~11.5:1 (after subtracting the background), which is much higher than that for the NPs obtained by the routine NSL and releasing process (1:3.6)(Song and Elsayed-Ali 2010). Clearly, the number of the small debris caused by the sonication is greatly reduced using the modified NSL and releasing process. The modified NSL process favors the formation of uniform Ag NPs with rounded tips with significant reduction in Ag debris, as shown in Fig. 15. In addition, 1-OT and 11-MUA can be substituted by the combination of 1-BT and TP, or MCH and MUA if more water-soluble NPs are desired.

Clearly, Ag NPs with controlled shapes and reduced defect density can be fabricated by a modified NSL process. Upon dislodging these NPs into a solution, they retain their shapes significantly better than NPs produced by routine NSL. Thus, aqueous phase Ag NPs with relatively uniform size and shape distribution can be fabricated. The UV-vis absorption spectra for surface confined NPs show two distinct absorption peaks (Figure 16a), comparing with those with rounded tips (Figure 16b). After surface modification, the central wavelengths of the two absorption peaks blue shifted and showed reduced intensities. The aqueous phased Ag NPs produced by the modified NSL method show a main peak and another peak with very low intensity attributed mainly to small debris produced during the dislodging process. The noticeable reduction in the intensity of the short wavelength peak for the modified NSL method compared to the routine method is due to the significant reduction in Ag debris. TEM images show that the uniformity of Ag NPs can be improved significantly by the modified NSL and releasing processes.

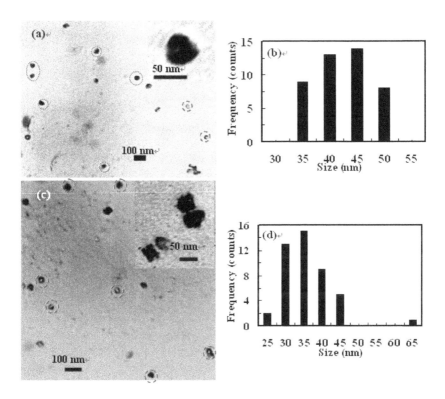

Fig. 15. TEM images of the aqueous phase Ag NPs after surface modification by thiol compounds and dislodging from the glass substrate. (a) Triangle Ag NPs with rounded tips. Dashed circles: Ag NPs with rounded tips; dotted circle: Ag NPs with slightly rounded tips. (b) Histogram of triangular shaped Ag NPs with rounded tips based on 45 NPs giving a mean size of 39.6±4.9 nm. (c) Quadrilateral and pentagon shaped Ag NPs. Dashed circles: some typical Ag NPs with quadrilateral shapes. (d) Histogram of quadrilateral and pentagon shaped Ag NPs based on 45 NPs giving a mean size of 33.9±6.8 nm. (Reprinted from Song et al., Appl. Surf. Sci. 2010 256, (20), 5961, Figure 4. Copyright (2010) Elsevier.)

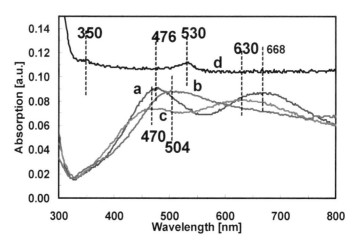

Fig. 16. UV-vis optical absorption of Ag NPs. (a) Surface confined Ag NPs before tip rounding and surface modification. (b) Surface confined Ag NPs after tip rounding (c) Surface confined Ag NPs with rounded tips after surface modification with thiol. (d) Aqueous phase Ag NPs after releasing the surface confined NPs into water. (Adapted from Song et al., Appl. Surf. Sci. 2010 256, (20), 5961, Figure 5. Copyright (2010) Elsevier.)

7. Perspective for the NSL in the controlled fabrication of nanomaterials

The great progress in controlled synthesis/fabrication of noble metal NPs by NSL, and the increase in the experimental and theoretical achievements in control of their size, shape, surface morphology and 3-dimensional space orientation dependent physicochemical properties and functions suggest expanding application in many fields because of the potential for essential breakthroughs by researchers and engineers for more advanced applications of NSL. Particularly, the developed multi-step angle resolved NSL and the modified NSL incorporated with suitable post-treatments have enabled us to obtain uniform surface-confined overlapped and nano-gapped nanostructures, the tip-rounded triangular nanoprisms, the square-shaped and the trapezoidal nanoprisms, besides the common triangular nanoprisms. Besides the marvelous progresses in the surface-confined nanostructures fabrication, a modified NSL process has also been developed to dislodge these uniform nanomaterials into the desired solvents (e.g. water, ethanol) without any obvious agglomeration as in the solution-phased nanocolloids synthesis.

Progresses in the incorporation of NSL with other LIGA techniques have shown that the suitability and ability in the architecture and interspacing controlled fabrication of NPs and nanoarrays. Their applications can thus be expanded extremely. When the multi-hierarchy arrayed micro windows fabricated by the traditional UV-LIGA process is joined in NSL, one powerful method for single nanoparticle identification will be born, resulting in the possibility of the precise investigation of the 3D morphology dependent LSPR of nanoparticles and LSPR coupling in nanoarrays. By collaboration with UV-LIGA microfabrication, uniform noble metal nanoparticles or nanoarrays can be fabricated into the targeted micro channels, leading to a much sensitive optical biosensing system after their surfaces are modified by the traditional functionalization process. By combination of PAA-LIGA and NSL, the possibility for building hierarchically ordered multi-segment nanowires or nanorods will be realized conveniently.

Summarizing from the recent progresses and discussion on NSL presented in this chapter, four main researches thrust that includes several active and challenging topics may form the primary research focuses and directions in this particular field. One is the fabrication technique development for the formation of monolayer of nanospheres with uniform area as large as several centimeter squares, which founds the basis of NSL. Another is the convenient and practical process in the releasing of these surface confined nanomaterials into solvent with perfectly retained 3D morphologies, which is still challenging but a desired alternative to obtain the uniform nanomateirals besides the well-developed wet chemical process. The third is the advanced incorporation of NSL with other fabrication techniques besides LIGA processes for the building more complex 3D hierarchically ordered nanostructures, which will definitely make a breakthrough in the nanoscale device and assemble development. The fourth may be the fabrication of tunable hetero-structure-composition nanocomposites, such as sandwich discs or multi-layer nanostructures, which will produce hetero-nanostructures with multi-functions (e.g. magntic, optical, electronic, etc). Consequently, outcomes of these challenging researches will result in the discovery of many exciting and versatile techniques for nanomaterials fabrication, and theoretical breakthrough in their novel physicochemical properties and for advanced applications.

8. Acknowledgement

The author appreciates the support from the basic research Vision Funds (YWF-11-03-Q-002) and, Chinese Scholarship Council (File No. 2010307428) NSFC (Grant No. 50971010).

9. References

Ahmadi, T. S.; Wang, Z. L.; et al. (1996). "Shape-Controlled Synthesis of Colloidal Platinum Nanoparticles." Science 272: 1924-1927.

Amanda, H. J.; Zhao, J.; et al. (2005). "Solution-Phase, Triangular Ag Nanotriangles Fabricated by Nanosphere Lithography." J. Phys. Chem. B 109(22): 11158-62.

Brayner, R. (2008). "The toxiccological impact of nanoparticles." Nano Today 3(1-2): 48-55.

Chong, M. A. S.; Zheng,Y. B.; et al. (2006). "Combinational template-assisted fabrication of hierarchically ordered nanowires arrays on substrates for device applications." Appl. Phys. Lett. 89: 233104-1-3.

Daniel, M.-C. and Astruc, D. (2004). "Gold Nanoparticles: Assembly, Supramolecular Chemistry, Quantum-Size-Related Properties, and Applications toward Biology, Catalysis, and Nanotechnology." Chem. Rev. 104: 293-346.

Dhar, P.; Cao, Y.; et al. (2007). "Autonomously Moving Local Nanoprobes in Heterogeneous Magnetic Fields." J. Phys. Chem. C 111: 3607-3613.

Edwards, H. W. and Petersen, R. P. (1936). "Reflectivity of evaporated silver films." Phys. Rev. 9: 871.

Erhardt, D. (2003). Materials conservation: Not-so-new technology Nature Materials 2: 509-510.

Haes, A. J.; Zou, S.; et al. (2004). "A Nanoscale Optical Biosensor: The Long Range Distance Dependence of the Localized Surface Plasmon Resonance of Noble Metal Nanoparticles." J. Phys. Chem. B 108: 109-116.

Hammond, C. R. (2000). The Elements, in Handbook of Chemistry and Physics 81th edition, CRC press.

Haynes, C. L.; Mcfarland, A. D.; et al. (2003). "Nanoparticle Optics: the Importance of Radiative Dipole Coupling in Two-Dimensional Nanoaprticle Arrays." J. Phys. Chem. B 107: 7337-7342.

Haynes, C. L. and van Duyne, R. P. (2001). "Nanosphere Lithography: A Versatile Nanfabrication Tool for Size Dependent Nanoparticle Optics." J. Phys. Chem. B 105: 5599.

Hong, B. H.; Bae, S. C. et al. (2001). "Ultrathin single-crystalline silver nanowire arrays formed in an ambient solution phase." Science 294: 348-351.

http://en.wikipedia.org/wiki/Silver.

Hulteen, J. C.; Treichel,D. A.; et al. (1999). "Nanosphere Lithography: Size-tunable silver nanosparticle and surface cluster arrays." J. Phys. Chem. B 103: 3854-3863.

Jain, P. K.; Huang, X.; et al. (2007). "Review of some interesting surface plasmon resonance-enhanced properties of noble metals nanoparticles and their applicaitons to biosystems." Plasmonics 2: 107-118.

Jensen, T. R.; Duval Malinsky, M.; et al. (2000). "Nanosphere Lithography: tunable localized surface plasmon resonance spectra of silver nanoparticles." J. Phys. Chem. B 104: 10549-10556.

Jin, R.; Cao,Y. W.; et al. (2001). "Photoinduced conversion of silver nanospheres to nanoprisms." Science 294: 1901-1904.

Kelly, K. L.; Coronado, E.; et al. (2003). "The optical proeprties of metal nanoparticles: the influence of size, shape and dielectric environment." J. Phys. Chem. B 107(3): 668-677.

Kreibig, U. and Vollmer, M. (1995). Optical Properties of Metal Clusters. Berlin, Springer.

Lee, S. J.; Morrill, A. R.; et al. (2006). "Hot spots in silver nanowire bundles for surface-enhanced Raman spectroscopy." J. Am. Chem. Soc. 128: 2200-2201.

Lombardi, I.; Cavallotti, P. L.; et al. (2007). "Template assisted deposition of Agnanoparticle arrays for surface-enhanced Raman scattering applications." Sensors and Actuators B: Chemical 125: 353-356.

Malinsky, M. D.; Kelly, K. L.; et al. (2001). "Chain length Dependence and sensing capabilities of the localized surface plasmon resonance of silver nanoparticles chemically modified with alkanethiol self-assembled monolayers." J. Am. Chem. Soc. 123: 1471-1482.

Maneerung, T.; Tokura, S.; et al. (2008). "Impregnation of silver nanoparticles into bacterial cellulose for antimicrobial wound dressing." Carbohydrate Polymers 72: 43-51.

Meng, G.; Jung,Y. J.; et al. (2005). "Controlled fabrication of hierarchically branched nanopores, nanotubes and nanowires." PNAS 102(20): 7074-7078.

Mock, J. J.; Barbic, M.; et al. (2002). "Shape effects in plasmon resonance of individual colloidal silver nanoparticles." J. Chem. Phys. 116(15): 6755-6759.

Mohl, M.; Kumar, A.; et al. (2010). "Synthesis of catalytic porous metallic nanorods by galvanic exchange reaction." J. Phys. Chem. C 114: 389-393.

Noguez, C. (2007). "Surface Plasmons on Metal Nanoparticles: the Influence of Shape and Physical Environment." J. Phys. Chem. C 111(10): 3806-3819.

Ozbay, E. (2006). "Plasmonics:Merging Photonics and Electronics at Nanoscale Dimensions." Science 311: 189-193.

Pan, S. L.; Zeng, D. D.; et al. (2000). "Preparation of ordered array of nanoscopic gold rods by template method and its optical properties." Appl. Phys. A 70: 637-640.

Percival, S. L.; Bowler, P. G.; et al. (2005). "Bacterial resistance to silver in wound care." Journal of Hospital Infection 60: 1-7.

Prasad, P. N. (2004). Nanophotonics. Hoboken, New Jersey, John Wiley & Sons, Inc.

Schwartzberg, A. M. and Zhang J. Z. (2008). "Novel Optical Properties and Energing Applications of Metal Nanostructures." J. Phy. Chem. C 112(28): 10324-10337.

Shen, Y.; Friend, C. S.; et al. (2000). "Nanophotonics: Interaction, Materials and application." J. Phys. Chem. B 104: 7577-7587.

Song, Y. (2009). Fabrication of high throughput biosensors based on single Nanoparticles and Nanoparticle arrays, China Patent, Appl. No. CN 200910085973.9.

Song, Y. (2010). "Fabrication of Multi-level 3-Dimension Microstructures by Phase Inversion Process." Nano-Micro Letters 2(2): 95-100.

Song, Y. and Elsayed-Ali, H. E. (2010). "Aqueous Phase Ag Nanoparticles with Controlled Shapes Fabricated by a Modified Nanosphere Lithography and their Optical Properties." Appl. Surf. Sci. 256(20): 5961-5967.

Song , Y.; Henry, L. L.; et al. (2009). "Stable Cobalt Amorphous Nanoparticles Formed by an In-situ Rapid Cooling Microfluidic Process " Langmuir 25 (17): 10209-10217.

Song , Y.; Jin, P.; et al. (2010). "Microfluidic Synthesis of Fe Nanoparticles." Mater. Lett. 64: 1789-1792.

Song, Y.; Sun, S.; et al. (2010). "Synthesis of Worm and Chain-like Nanoparticles by a Microfluidic Reactor Process." J. Nanopart. Res. 12: 2689-2697.

Song , Y.; Zhang, Z.; et al. (2011). "Identification of Single Nanoparticles." Nanoscale 3: 31-44.

Su, K.-H.; Wei, Q.-H.; et al. (2003). "Interparticle Coupling Effects on Plasmon Resonances of Nanogold Particles." Nano lett. 3(8): 1087-1090.

Vo-Dinh, T.; Wang, H.-N.; et al. (2009). "Plasmonic nanoprobes for SERS biosensing and bioimaging." J. Biophoton.: 1-14.

Xu, C.-l.; Zhang, L.; et al. (2005). "Well-dispersed gold nanowire suspension for assembly applicaiton." Appl. Surf. Sci. 252: 1182-1186.

Xu, Q., Meng, G.; et al. (2009). "A Generic Approach to Desired Metallic Nanowires Inside Native Porous Alumina Template via Redox Reaction." Chem. Mater. 21: 2397-2402.

Yang, Y.; Matsubara, S.; et al. (2007). "Solvothermal synthesis of multiple shapes of silver nanoparticles and their SERS properties." J. Phys. Chem. C 111: 9095-9104.

Zhang, J.; Li, X. et al. (2005). "Surface enhanced raman scattering effects of silver colloids with different shapes." J. Phys. Chem. B 109: 12544-12548.

Zhang, X.; Whitney, A. V.; et al. (2006). "Advances in Contemporary Nanosphere Lithographic Techniques." J. Nanosci. Nanotech. 6: 1920-1934.

Zhao, L. L.; Kelly, K. L.; et al. (2003). "The extinction spectra of Ag nanoparticle arrays: influence of array structure on plasmon resonance wavelength and widths." J. Phys. Chem. B 107: 7343-7350.

Zhou, Q.; Qian, G.; et al. (2008). "Two-dimentional assembly of silver nanoparticles for catalytic reduction of 4-nitroaniline." Thin Solid Films 516: 953-956.

Zhu, S.; Li, F.; et al. (2008). "A localized surface plasmon resonance nanosensor based on rhombic Ag nanoparticle array." Sensors and Actuators B: Chemical 134: 193-198.

8

Nanolithography Study Using Scanning Probe Microscope

S. Sadegh Hassani and H. R. Aghabozorg
Research Institute of Petroleum Industry
Iran

1. Introduction

Recently, the interest to design useful nanostructures in science and technology has rapidly increased and these technologies will be superior for the fabrication of nanostructures (Iwanaga & Darling, 2005; Martin et al., 2005; Sadegh Hassani et al., 2010). The patterning of material in this scale is one of the great importances for future lithography in order to attain higher integration density for semiconductor devices (Sadegh Hassani & Sobat, 2011).

Conventional lithography techniques, i.e., those divided to optical and electron beam lithography are either cost-intensive or unsuitable to handle the large variety of organic and biological systems available in nanotechnology. Hence, the various driving forces have been considered for development of nanofabrication techniques (Geissler & Xia, 2004; Quate, 1997; Sadegh Hassani & Sobat, 2011).

Applying of these techniques has started approximately since 1990 and it has given rise to the establishment of different nanolithography methods, which one of the most important method is scanning probe based lithography. An interesting way of performing nanometer pattern is direct scratching of a sample surface mechanically by a probe. The controlled patterning of nanometer scale features with the scanning probe microscope (SPM) is known as scanning probe lithography (SPL) (Irmer et al., 1998). Many reports have been presented about various lithographic methods by this technique (Garcia, 2004; Garcia, 2006). SPL would also be ideal for evaluation of mechanical characteristic of surfaces.

Scanning probe microscopy, such as scanning tunneling microscopy (STM) and atomic force microscopy (AFM), has become a standard technique for obtaining topographical images of surface with atomic resolution (Hyon et al., 1999). In addition, it may be used to study friction force, surface adhesion and modifying a sample surface (Sundararajan & Bhushan, 2000; Burnham et al., 1991; Aime et al., 1994; Sadegh Hassani & Ebrahimpoor Ziaie, 2006; Ebrahimpoor Ziaie et al., 2008). Manipulating surfaces, creating atomic assembly, fabricating chemical patterns and characterizing various mechanical properties of materials in nanometer regime are enabled by this technique (Hyon et al., 1999; Sadegh Hassani & Sobat, 2011; Bouchiat & Esteve, 1996).

Nanolithography with AFM is also a tool to fabricate nanometer-scale structures with at least one lateral dimension between the size of an individual atom and approximately 100 nm on silicon or other surfaces (Wilder & Quate, 1998). This technique is used during the fabrication of leading edge semiconductor integrated circuits (Sugimura & Nakagiri, 1997) or nanoelectromechanical systems (NEMS). This method is not restricted to conductive materials (Fonseca Filho et al., 2004). The advantages of this technique are high resolution and alignment accuracy, which could not be achieved by conventional lithographic techniques (Sheehan & Whitman, 2002; Martin et al., 2005). Moreover, the AFM nanolithography technique takes advantages of the ability to move a probe over the sample in a controllable way (Samori, 2005). Nanolithography using AFM can be done in various modes (Jones et al., 2006): chemical and molecular patterning (DPN), mechanical patterning by scratching or nanoindentation, local heating, voltage bias application and manipulation of nanostructures. Most popular AFM lithographic techniques are resist film lithography (Li et al., 1997) and lithography by oxidation (Sheglov et al., 2005; Sugimura et al., 1993; Sadegh Hassani et al., 2008a; Dubois & Bubbendroff, 1999; Avouris et al., 1998; Snow et al., 1999; Lemeshko et al., 2005; Avouris et al., 1997). The atomic force microscope has also become an increasing popular tool for manipulating thin films of many different types of materials. Lithography techniques can be carried out on the film of polymers such as polymethylmethaacrylate (PMMA), chloromethyl phenyltrichlorosilan (CMPTS), polyethylene (PE) and others (Lee et al., 1997; Sadegh Hassani et al., 2008b, Yoshimura et al., 1993; Chen et al., 1999; Huang et al., 2001). This capability can potentially be extended to evaluate nano-scale material response to indentation and would be ideal for evaluation of mechanical characteristic of surfaces (Burnham & Colton, 1989; Hues et al., 1994; Sadegh Hassani et al., 2008b).

To apply force optimally for making nano scratches, we require to understand the underlying behavior and parameter that control it, a tip which is optimized for applying force under the experimental conditions and scanning techniques which allows one to use these tips and retain desired properties (Yasin et al., 2005; Sadegh Hassani et al., 2008a; 2010). Some factors such as resolution, accuracy of alignment and reproducibility are important in this way. By reducing wear of AFM tip and controlling variables such as applying force, scan speed and environment, it can be systematically calibrated the size of features that is written by AFM tip. So, the reproducibility of issues can be controlled.

In this chapter, it is focused on the use of lithography process to build the desired nanostructures and nanolithography on surface of different substrates by AFM. Creating the scratches on various surfaces by silicon nitride and diamond tips using contact mode is discussed. For scratching, the mechanical action of the tip as a sharp pointed tool in order to produce fine scratches is used (Notargiacomo et al., 1999; Sadegh Hassani et al., 2010). The direct scratching is possible with high precision but low quality results are obtained due to probe wear during lithographic process.

Silicon nitride cantilever tip with average spring constant is used to investigate soft surfaces including poly methyl methacrylate (PMMA) (LG-IH 830) thin film coated on the silicon and glass substrates (Sadegh Hassani et al., 2008a). A diamond cantilever tip with high spring constant is used for hard surfaces including highly-oriented pyrolytic graphite (HOPG) and polyethylene substrate (Sadegh Hassani et al., 2010). Since its hardness is much more than silicon nitride, the direct formation of nanoscratches could easily be achieved.

Effects of applied normal force, time of applying pressure, speed and number of scratching cycles on the geometry and depth of scratches are studied. This study shows that there is a critical tip force to remove material from various surfaces (Sadegh Hassani et al., 2008a; 2008b; 2010).

2. Force lithography

An interesting way of performing nanometer pattern is force lithography which based on direct mechanical impact produced by a sharp probe on the sample surface (Lyuksyutov et al., 2003; Park et al., 2000; Sadegh Hassani et al., 2008a). The probe tip pressure on the surface is sufficient to cause plastic deformation of the substrate surface. This type of modification has been used in nanoelectronics, nanotechnology, material science, etc. It enables the fabrication of electronic components with active areas of nanometer scale, super dense information recording and study of mechanical properties of material.

In force lithography no bias voltage is required to produce nanostructures. The nanostructure formation normally occurs as a result of AFM tip motion above the polymer surface with set point magnitude constraining the tip to come closer to the surface (Lyuksyutov et al., 2004; Sadegh Hassani et al., 2008a; 2008b; 2010). In order to apply sufficient normal load to reach plastic deformation of surface, a three-side pyramidal single crystalline diamond tip or another tip with high spring constant is used and pressed against a desired surface (Santinacci et al., 2003; Sadegh Hassani et al., 2010). Much higher forces are achieved by accordingly increasing the applied voltage to piezo-scanner. By scanning the sample in the X or Y direction at various conditions (such as different scanning velocity and number of cycles) grooves are created. However, the protrusions along the edges are formed, which indicates clearly stress deformation during the scratching process (Santinacci et al., 2005).

It is shown that by applying a little force (several μN), removing an amount of material from a metal or polymer film is possible (Bruckl et al., 1997).

Use of cantilevers with high spring constant could apply the desired amount of force without large bending. When tip move toward the substrate or reverse direction, up or down bending of cantilever occurs, respectively. Since an angle of about 10° is typically set between the cantilever and the substrate (see Fig.1. b), this bending influence the tip-substrate interaction, so the geometry and size of scratches are affected in this way. However, increase of applied force cause cumulating of material along or at the end of the grooves. This deformity is occurred because of cantilever bending at the start point of moving tip through the surface. In this way cantilever reach the desired force to create scratch. (Notargiacomo et al., 1999; Sadegh Hassani & Sobat, 2011).

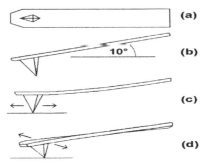

Fig. 1. Typical silicon cantilever with pyramidal tip: (a) upper view; (b) lateral view showing the 10° angle formed with the substrate surface; (c) cantilever bending and (d) torsion (Notargiacomo et al., 1999).

Resolution will be major challenge in lithographic fabrication and the limiting factor for resolution is the tip quality. Sharp silicon tips deliver brilliant and reproducible results. To even further achieve the fine lithographic structure, electron beam deposited tips (EBD tips) can be additionally sharpened in oxygen plasma (Wendel et al., 1995).

Wearing of probe led to low-quality results and reduced the repeatability of produced scratches. Indeed, by using the same tip at another experiment, the sample surface could experience two completely different values of pressure, because the amount of produced pressure depends on the shape of tip (Hu et al., 1998). To decrease wearing of probe, a soft resist polymer film (usually PMMA film) is coated on the surface. On the other hand, the roughness of surface is very important to take high quality scratches. Observations show that surface roughness is strongly influenced by its thickness as while; the surface roughness increases with the increase of the thickness. So, to perform the lithography process, the smoothest surface has to be chosen (Yasin et al., 2005; Fonseca et al., 2004).

Studies show that in the case of AFM, the possibility of directly machining a surface has been explored in two ways, i.e. by either using a static approach in which the microscope is operated in conventional contact mode (Magno & Bennett, 1997; Sumomogi et al., 1995) or using a dynamic approach in which the microscope is operated in the tapping mode (Heyde et al., 2001; Davis et al., 2003). Usually the lithography developed using both static and dynamic approaches are employed to pattern a resist layer, subsequently the patterned layer is used as an etch mark. Both techniques are giving lithography resolution of the order of tens of nanometer (Wendel et al., 1994; Quate, 1997).

An advantage of the vibration in the tapping mode is that very small lateral forces stress the tips, resulting in very slow tip degradation (Wendel et al., 1996).

3. Force curve

In lithographic experiments, it is often critical to know the tip pressure on the desired sample. To estimate the pressure corresponding to a specified level of the probe impact, the force created by the probe has to be determined. It can be calculated from force spectroscopy data.

The normal force between tip and sample is estimated from cantilever deflection (nA) curve plotted against Z-displacement of the cantilever and converting this curve to Force-Distance curve (Vanlandingham, 1997; Yeung et al., 2004; Carallini et al., 2003; Santinacci et al., 2005; Sadegh Hassani et al., 2008a; Argento & French, 1996). To take the force curve, to avoid punching surface, it is essential that the sample has a rigid surface such as silicon or polycrystalline substrate. By performing spectroscopy in a point, force curve is obtained.

The conversion factor for converting nA to nm was obtained from the slope of the linear portion of the deflection–distance curve. There was also one conversion needed for the X–axis values. The change in piezo height, which has been used for the distance between the tip and the sample, was corrected for the deflection of the cantilever by subtracting the cantilever deflection from the piezo height.

On the other hand, there are two measurements required to convert photo detector signal into a quantitative value of force. The first stage is to calibrate the distance that the cantilever actually deflects for a certain measured changes in photo detector voltage. This value depends on type of cantilever and the optical path of the AFM detection laser. When every cantilever is mounted in the instrument, this value will be slightly different. Once the deflection of cantilever is known as a distance Z, the spring constant k, is needed to convert this value into a force F, using Hook's law (F = k × ΔZ) (Heimberg & Zandbergen, 2004; Ebrahimpoor Ziaie et al., 2005; Carpick & Salmern, 1997; Sadegh Hassani & Ebrahimpoor Ziaie, 2006).

4. Nanolithography on various substrates

In some reported experiments, a commercial scanning probe microscope (Solver P47H, NT-MDT Company), operated in AFM contact and noncontact modes, equipped with (NSG11) and (DCP20) cantilevers were used to perform the lithography of desired surfaces (Sadegh Hassani et al., 2008a; 2008b; 2010). The NSG11 cantilever made of silicon nitride had a rectangular shape, and its length, width and thickness were 100 ± 15 µm, 35 ± 3 µm and 1.7–2.3 µm, respectively. Its normal bending constant measured by supplier was 11.5 nN/nm. Another cantilever which was used in this process was DCP20 Cantilever made of diamond with the length, width and thickness of 90±5 µm, 60±3 µm and 1.7-2.3 µm, respectively. The normal bending constant measured by supplier was 48 nN/nm.

These two types of cantilever were selected to reach deformation of different types of surfaces and also for obtaining good images of scratches. These experiments were designed to fabricate scratches on the various surfaces with the different rigidity.

The lithography process was executed with the use of lithography menu supported by the microscope software. The AFM tip was brought into contact with the sample surface using the smallest force possible to minimize any undesired surface modification. An image of surface was prepared in order to choose a suitable surface free of defects for lithography; then the nanolithography process was executed under various specific and controlled conditions to analyze the effect of lithography important factors on the shape of scratches.

For studying force effect, the force was increased by applying a higher voltage to the piezo-scanner in order to reach the cantilever deflection (ΔZ) corresponding to the force (F) range where plastic deformation of polymeric surface occurred. (Santinacci et al., 2003; Notargiacomo et al., 1999; Sadegh Hassani et al., 2008 b). Scratches were made in Y direction on various substrates in different conditions (Sadegh Hassani et al., 2008a; 2008b; 2010), so in this way the influences of applied normal force, scanning velocity, time of applying pressure and number of scratching cycles were investigated. Finally surface was scanned by atomic force microscope in non-contact mode to observe and evaluate the shape and depths of scratches. If the contact mode had been chosen to image the scratches, the surface of substrates would have probably been damaged.

4.1 Nanolithography on PMMA thin films

Sadegh Hassani et al. (2008a; 2008b; 2010) reported lithography performance on PMMA thin films. In this regard, soft thin films of PMMA polymer on the silicon and glass substrates were prepared. For making PMMA (LG-IH830) thin film on silicon and glass substrates, these substrates were washed and sonicated in acetone/ethanol (50–50 % vol.) for 15 minutes at room temperature. Then a very small amount of diluted PMMA/CHCl$_3$ solution was coated over the silicon and glass surfaces using spin coater with 6000 rpm for 30 seconds. The coated substrates were dried in an oven at 130 °C for 30 minutes. The thickness of these coated layers was ~150 nm, measured by atomic force microscope.

In order to choose suitable area for nanolithography process, the topography images and the roughness of the surfaces of PMMA thin films were investigated with the AFM (Sadegh Hassani et al., 2008a). It was reported (Notargiacomo et al., 1999) that "a high value of the surface roughness could produce unwanted features and inhomogeneous results during patterning". At the first step, it was necessary to evaluate the substrate surface after cleaning. In Figure 2 the evolution of the topography images and profiles of the silicon surface after cleaning and PMMA thin films are presented. It is seen that the roughness of PMMA thin film is low and its surface profile is appropriate for lithography. An accurate study was performed on the samples in order to find the optimum patterning conditions for the PMMA film.

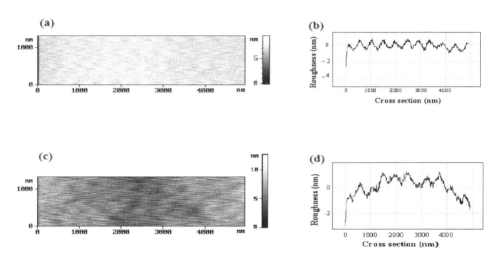

Fig. 2. The evolution of the topography images and profiles of (a & b) silicon surface after cleaning and (c & d) PMMA thin film (Sadegh Hassani et al., 2008a).

For PMMA coated on silicon substrate, scratches were performed using the NSG 11 tip. The main factor in pattern formation was the magnitude of the force applied to the sample. The influence of the applied normal force on the scratches created on the PMMA film coated the silicon substrate had been investigated. In Figure 3 (a-e) some of surface profiles of nanoscratches are presented, which are formed with the constant scanning velocity of 140 nm/s, number of scratching cycle of 10 within 25 ms at various forces (2350, 2700, 3050, 3400, and 3510 nN).

These profiles indicate that the increase of applied normal force, leads to the deeper scratches. The scratches are V-shape; however protrusions are visible along some of the scratches indicating the presence of permanent deformation. It was found that the optimum value for applied normal force was about 3050 nN. The scratch made by this force is shown in Figure 4. In Figure 5, the scratch depths are plotted as a function of the applied normal force. As expected, the scratch size increases with increasing the force load. The depth varies from 4 to 32 nm by increasing force load from 1300 to 3510 nN. However, Notargiacomo et al. claimed (1999) that as the applied force increases, curved cuts ("tails") become visible at the ends of the lines. It has to be mentioned that due to the convolution effect of the tip and substrate topography, the scratch depth may appear smaller by AFM imaging than their actual size (Sadegh Hassani et al., 2008a).

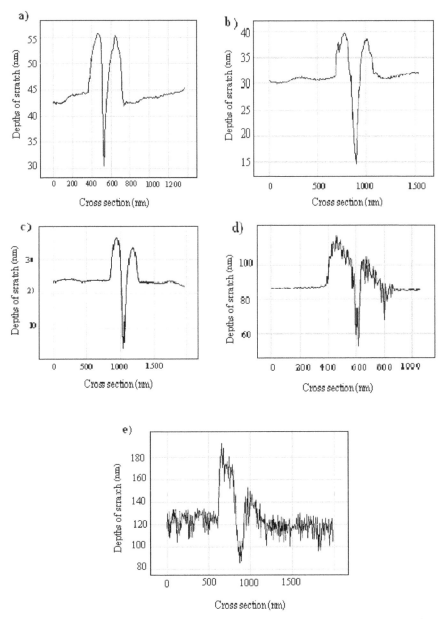

Fig. 3. (a-e) Surface profiles for scratched PMMA film (Sadegh Hassani et al., 2008a) (N= 10 cycles, T=25 ms, V=140 nm/s and F is equal to (a) 2350 nN, (b) 2700 nN, (c) 3050 nN, (d) 3400 nN and (e) 3510 nN).

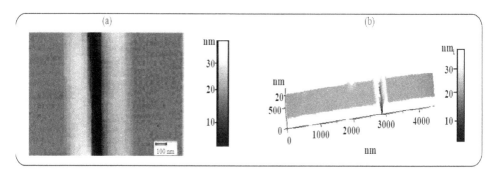

Fig. 4. Topography images of the scratch performed on the PMMA /silicon (Sadegh Hassani et al., 2008a; 2008b; 2010) at N= 10 cycles, T=25 ms, V=140 nm/s and F= 3050 nN. (a) Two dimensional image and (b) Three dimensional image

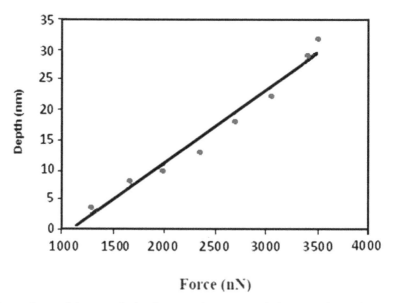

Force (nN)

Fig. 5. Dependence of the scratch depth, created on PMMA/silicon, to the applied normal force (Sadegh Hassani et al., 2008a; 2008b; 2010). (The time of applying pressure, number of scratching cycle and scanning velocity are 25 ms, 10 and 140 nm/s, respectively.)

For PMMA coated glass substrate, scratches were performed with exerting various normal forces using NSG 11 tip. In Figure 6, the groove depths are plotted as a function of the applied normal force for PMMA on glass substrate. The most uniform scratches were achieved by applying 3,000 nN force load, while scanning velocity, number of scratching cycle and time of applying pressure were 1400 Å/s, 10 and 25 ms, respectively. Topography image of this scratch is shown in Figure 7. The uniformity of scratches on PMMA coated on silicon and glass is comparable. However, the depth of scratch on the PMMA/glass at the same conditions is more than that of on the PMMA/silicon.

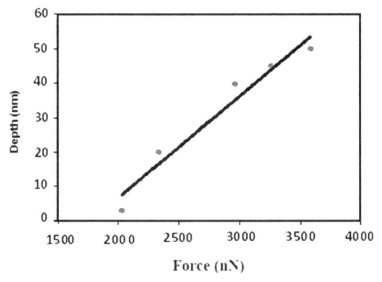

Fig. 6. Dependence of scratches depth, created on PMMA/glass (Sadegh Hassani et al., 2010), with the applied normal force while scanning velocity, number of scratching cycle and time of applying pressure were 140 nm/s, 10 and 25 ms, respectively.

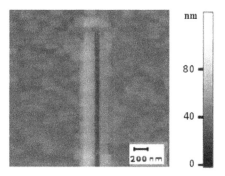

Fig. 7. Topography image of the scratch created on PMMA/glass (Sadegh Hassani et al., 2010), while the applied normal force, scanning velocity, number of scratching cycle and time of applying pressure were 3000 nN, 140 nm/s, 10 and 25 ms, respectively.

The influence of the number of scratching cycles was also investigated by scratching experiments. In Figure 8 dependence of the scratch depth to the number of cycles (N = 1, 5, 10, 15, 20, 25, and 30) in a constant applied normal force of 2350nN, scanning velocity of 140 nm/s in 25 ms is presented. This figure shows that the depth varies from 4 to 30 nm by increasing the number of cycles. As expected, the depths of scratches increase with N linearly. This linear relationship between depth and number of cycles confirms layer-by-layer removal mechanism (Sadegh Hassani et al., 2008a). This result is in agreement with that of obtained by Santinacci and coworkers (2003) for performing nanolithography on p-Si (100) substrate.

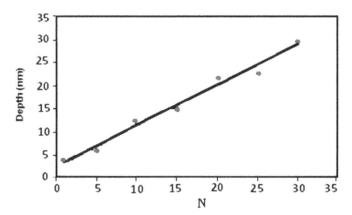

Fig. 8. Dependence of scratch depth created on PMMA/Si to the number of cycles (Sadegh Hassani et al., 2008a; 2008b) (The applied normal force, scanning velocity and time of applying pressure are 2350 nN, 140 nm/s, and 25 ms, respectively.)

The influence of the scanning velocity on the lithography pattern, taken at a normal force of 2350 nN is presented in Figure 9. As it is shown in this figure, the depth varies from 24 to 8 nm by decreasing scanning velocity from 140 to 540 nm/s. It is observed that the increase of the scanning velocity induces a decrease in the scratch depth. Thus, slower scans seem to generate higher pressure and as a result deeper scratch pattern are obtained. However, it could not be determined whether the depth decreases linearly or exponentially with the increase of the scanning velocity (see Fig. 9). To analyze the time effect, nanoindentations are performed on the PMMA surface. The indentation depth created on PMMA/Si depth as a function of time of applying pressure using constant force is presented in Figure 10. It can be seen that by increasing the time of applying pressure, indentation depth is increased. In the other words, the plastic deformation on the PMMA film is time dependent. In this case, to accumulate the tip-induced stress, dilation changes such as defects created or absorbed near the vicinity of the deformed region on the surface occur. This effect leads to an additional plastic deformation of the film.

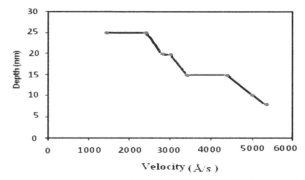

Fig. 9. The influence of the scanning velocity on the depth of the lithographed pattern created on PMMA/Si (Sadegh Hassani et al., 2008a) (The applied normal force, time of applying pressure and number of cycle are 3125 nN, 25 ms and 10, respectively.)

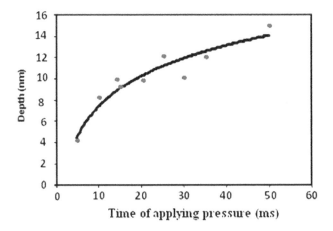

Fig. 10. The indentation depth created on PMMA/Si as a function of the time of applying pressure (Sadegh Hassani et al., 2008a; 2008b) (The applied normal force, scanning velocity and number of cycle are 2350 nN, 140 nm/s and 10, respectively.)

4.2 Nanolithography on polyethylene substrate

This experiment was performed on the polyethylene (PE) substrate (Sadegh Hassani et al., 2008b; 2010). Polyethylene surface was cleaned by washing and sonicating in acetone-ethanol (50-50%Vol.) for 15 minutes at room temperature. This substrate was more inflexible than PMMA thin layer, so performing any modification over PE needed more rigid cantilever tip. The results verified this comment. Scratches were just made by maximum amount of force load, which was equal to 4 µN for NSG11 cantilever tip that was the threshold of force for modifying the PE substrate. The investigation of force effect on the PE substrate was continued by DCP20 cantilever with diamond tip. The force load created by this tip was sufficient to make modification on the PE substrate because of higher spring constant.

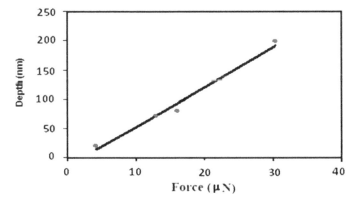

Fig. 11. Dependence of scratches depth with the applied normal force on the PE substrate (Sadegh Hassani et al., 2008b; 2010), while scanning velocity, number of scratching cycle and time of applying pressure are 140 nm/s, 20 and 50 ms, respectively.

A topography image of nanoscratche made on the PE showed that the best quality of scratch was obtained by applying of a 4μN force load. The obtained results showed that the uniformity of the scratches reduced by increasing the force load. Accumulation the vicinity of scratches was occurred, because increasing the applied force induced additional plastic deformation. Figure 11 shows the linear increase of scratches depth on the PE substrate as a function of applied normal force. Meanwhile, increase of applied force caused cumulating of material at the start and end point of the grooves. This deformity was occurred because of cantilever bending at the start point of moving tip through the surface (Notargiacomo et al., 1999; Sadegh Hassani et al., 2088b).

In Figures 12 and 13, the indentation depths are plotted as a function of the time of applying pressure and number of cycles for PE substrate, respectively. Figure 12 shows that the dependence of depth to time is not quite linear. This result was also reported by Santinacci et al. (2005). Figure 13 shows the linear increase of scratches depth with number of cycles, as expected (Santinacci et al., 2003; Santinacci et al., 2005; Sundararajan & Bhushan, 1998).

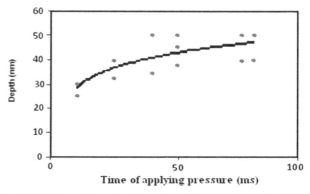

Fig. 12. Indentation depth dependence to time of applying pressure for PE substrate (Sadegh Hassani et al., 2008b), while applied normal force, scanning velocity and number of scratching cycle are 4μN, 140 nm/s, 10, respectively.

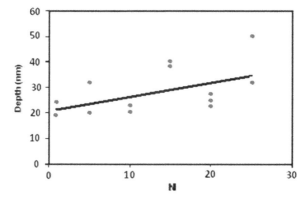

Fig. 13. Indentation depth dependence to number of scratching cycle for PE substrate (Sadegh Hassani et al., 2008b), while applied normal force, scanning velocity and time of applying pressure are 4μN, 140 nm/s, 25 ms, respectively.

According to Figures 12 and 13, repeatability of results for PE substrate is less than PMMA layer. It refers to roughness and flexibility of PMMA thin film. In the other words, making scratch over very uniform and flexible PMMA thin layer is much more successful than PE substrate.

4.3 Nanolithography on HOPG

This experiment was performed on the HOPG substrate as a completely hard surface and by exerting applied normal force ranging from 5.5 to 50.5 µN (Sadegh Hassani et al., 2010). Moreover, in comparison with PMMA thin films and polyethylene substrate, the scanning velocity for performing nanolithography on HOPG surface had to be increased which would led to wearing tip very fast. However, the time of applying pressure on HOPG was much less than those applied on PMMA thin films and polyethylene substrate. Hence, after some trial experiments, 10000 nm /s was chosen for scanning velocity on HOPG substrate at 1 ms. HOPG surface was cleaned using double-sided tape and removing one layer of it. The results of this experiment are obtained by DCP20 cantilever and are presented in Figure 14. Because HOPG surface was very uniform with very low roughness, shape of made scratches were completely V-form. Topography image and cross section of one of the scratches performed on the HOPG is shown in Figure 15. Meanwhile, increase of applied normal force caused cumulating of material at the start and end point of the grooves. This deformity occurred because of cantilever bending at the start point of moving tip through the surface (Wendel et al., 1996).

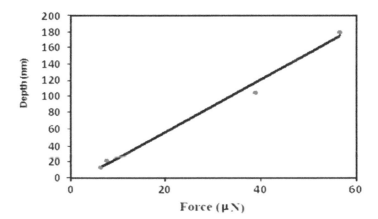

Fig. 14. Dependence of scratches depth to the applied normal force executed by DCP20 cantilever on the HOPG (Sadegh Hassani et al., 2010) substrate, while scanning velocity, number of scratching cycle and time of applying pressure are 10000 nm /s, 1 and 1 ms, respectively.

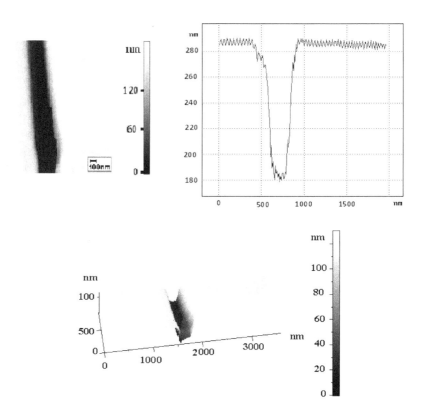

Fig. 15. Two and three-dimensional topography images and cross-section of one of the scratches performed on the HOPG (Sadegh Hassani et al., 2010) , while scanning velocity, number of scratching cycle, time of applying pressure and applied normal force are 10000 nm /s, 1, 1 ms and 50.4 µN, respectively.

5. Conclusion

This chapter is focused on the study of the Scanning Probe Lithography in a controlled way on various substrates. The load force produced by silicon nitride (NSG11) tip with average spring constant is sufficient to deform and make scratch on the PMMA thin film. The analysis of the roughness of the surface shows that the concept using a thin insulting film of PMMA on silicon and glass surfaces as a scratching mask can be successfully performed for nanopatterning. Drawing patterns are being controlled by the necessary parameters such as normal force, scanning velocity, time of applying pressure and number of scratching cycle. It is shown that the depth of the lithography mark increases linearly with the increase of the applied normal force. The uniformity of scratches on the PMMA coated on silicon and glass is comparable.

The load force applied by NSG11 tip is not sufficient to exert scratch on the hard surface and is disabled to perform any changes, so diamond tip with much higher spring constant is required. It must be mentioned that the minimum necessary force to modify the PE surface is about 4 μN that can be achieved by NSG11 tip and with maximum force load. Therefore, for investigation of the effect of higher forces, DCP20 tip is used for PE substrate. The experimental results show that depth of the lithography pattern increased with the increase of the applied normal force with a linear trend for all of the applied substrates. The increase of applied normal force caused accumulating of material at the start and end point of the grooves. This deformity occurred because of cantilever bending at the start point of the tip moving through the surface.

It is presented that the increase of the scanning velocity induces a decrease in the scratch depth. Thus, slower scans seem to generate higher pressure and as a result, deeper scratch pattern are obtained. However, it could not be determined whether the depth decreases linearly or exponentially with the increase of the scanning velocity.

The increase of the lithography depth with the loading time suggests that the plastic deformation on PMMA layer is time dependent. However, the results show that the dependence of depth to the loading time is not quite linear. By the tip-induced stress, dilation changes such as defects created or absorbed near the vicinity of the deformed region on the surface might occur, which would lead to an additional plastic deformation of the film.

It is shown that the depth of the lithography mark increases with the increase of the number of scratching cycle. The depths of scratches increase linearly with the number of scratching cycle.

Finally, due to the convolution effect of the tip and substrate topography, the scratch depth may appear smaller by AFM imaging than their actual size.

6. References

Aime, J. P.; Elkaakour, Z.; Odin, C.; Bouhacina, T.; Michel, D.; Curely, J. & Dautant, A. (1994). Comments on the use of the force mode in atomic force microscopy for polymer films. *Journal of Applied Physics*, 76, 754-762.

Argento, C. & French, R.H. (1996). Parametric tip model and force–distance relation for Hamaker constant determination from aromic force microscopy. *J. Appl. Phys.*,80 (11) 6081-6090.

Avouris, Ph.; Martel, R.; Hertel, T. & Sandstrom, R. (1998). AFM tip induced and current local oxidation of silicon and metals. *Appl. Phys. A* 66, S659-S667.

Avouris, Ph.; Hertel, T. & Martel, R. (1998). Atomic force microscope tip- induced local oxidation of silicon: kinetics, mechanism and nanofabrication. *Appl. Phis. Lett.* 71(2) 285-287.

Bouchiat, V. & Esteve, D. (1996). Lift-off lithography using an atomic force microscope, *Appl. Phys. Lett* ,69(20) 3098-3100.

Bruckl, H.; Rank, R.; Vinzelberg, H.; Monch, I.; Kretz, L. & Reiss, G. (1997). Observation of coulomb-blockade effects in AFM-machined tunnel junctions. *Surf. Interface Anal.* 25, 611-613.

Burnham, N. A. & Colton, R.J. (1989). Measuring the nanomechanical properties and surface forces of materials using an atomic force microscope. *Journal of Vacuum Science Technology*, A7, Issue 4, 2906-2913.

Burnham, N. A.; Colton, R. J. & Pollock, H. M.(1991). Interpretation issues in force microscopy. *Journal of Vacuum Science & Technology A: Vacuum, Surfaces, and Films*, Vol.9, Issue 4, 2548-2556.

Carallini, M.; Biscarini, F.; Leon. S.; Zerbetto, F.; Bottari, G. & Leigh, D.A. (2003). Information Storage using Supramolecular Surface Patterns. *Science* 299, 531.

Carpick, R. W. & Salmern, M. (1997). Scratching the surface: Fundamental investigations of tribology with atomic force. *Chem. Rev.* 97, 1163-1194.

Chen, J.; Read, M. A.; Raweltt, A. M. & Tour, J. M. (1999). Large on- off ratios negative differential resistance in a molecular electronic device. *Science,* 286, 1550-1552.

Davis, Z. J.; Abadal, G.; Hansen, O.; Borise, X.; Barniol, N.; Perez-Murano, F. & Boisen, A. (2003). AFM lithography of aluminum for fabrication of nanomechanical systems. *Ultramicroscopy* 97, 467-472.

Dubois, E. & Bubbendroff, J. L. (1999).Nanometer sacle lithography on silicon, titanium and PMMA resist using scanning probe microscopy. *Solid-state Electronics,* 43, 1085-1089.

Dienwiebel, M.; Verhoeven, G. S.; Pradeep, N.; Frenken, J.W.; Heimberg, J. A. & Zandbergen, H. W.,(2004). Superlubricity of graphite. *Phys. Rev. Lett.,* 92(12), (126101-1)-(126101-4).

Ebrahimpoor Ziaie, E.; Rachtchian, D. & Sadegh Hassani, S. (2008). Atomic force microscopy as a tool for comparing lubrication behavior of lubricants. *material science: An Indian journal,* 4(2), 111-115.

Ebrahimpoor Ziaie, E.; rachtchian, D. & Sadegh Hassani, S. (2005). Atomic Force Microscopy as a tool for comparing lubrication behavior of lubricants. *IV international conference on tribochemistry,* 3-5 october, Poland.

Fonseca Filho, H. D.; Mauricio, M. H. P.; Ponciano, C.R. & Prioli, R. (2004). Metal layer mask patterning by force microscopy lithography. *Material Science and Engineering* B., 112, 194.

Garcia, R.; Martinez R.V. & Martinez, J. (2006). Nano-chemistry and scanning probe nanolithographies. *Chem. Soc. Rev,* 35, 29-38.

Geissler, M. & Xia, Y. (2004). Patterning: Principles and Some New Development. *Adv. Mater.* 16, 1249-1269.

Heimberg, J. A. & Zandbergen, H. W. (2004). Superlubricity of graphite. *Physical Review Letters* 92 (12), 126101 (1) -126101.

Heyde, M.; Rademann, K.; Cappella, B.; Guess, M.; Strum, H.; Spanegenberg, T. & Niehus, H. (2001). Dynamic plowing nanolithography on polymethylmethacrylate using an atomic force microscope. *Rev. Sci. Instrum.* 72, 136-141.

Hu, S.; Hamidi; A.; Altmeyer; S.; Koster, T.; Spangenberg, B. & Kurz, H. (1998). Fabrication of silicon and metal nanowires and dots using mechanical atomic force lithography. *J. Vac. Sci. Technol.* B16, 2822-2824.

Huang, Y.; Duan, X.; Wie, Q. & Lieber, C. M. (2001). Directed assembly of one-dimensional nanostructures into functional networks. *Science,* 291, 630-633.

Hues, S. M.; Draper, C. F. & Colton, R.J., (1994). Measurement of Nanomechanical Properties of Metals Using the Atomic-Force Microscope. *J. of Vacuum Science Technology,* B12, 2211-2214.

Hyon, C. K.; Choi, S. C.; Hwang, S.W.; Ahn, D.; Kim, Y. & Kim, E. K. (1999). Direct nanometer scale patterning by the cantilever oscillation of an atomic force microscope. *Appl. Phys. Lett.,* Vol. 75, No. 2, 292–294.

Irmer, B.; Kehrle, M.; Lorenze, H. & Kothaus, J. P.,(1998). Nanolithography by non-contact AFM-induced local oxidation: fabrication of tunnelling barriers suitable for single-electron devices, *Semicond. Sci. Technol.,* 13, A79-A82 .

Iwanaga, Sh. & Darling, R.B. (2005). Stable and erasable patterning of vanadium pentoxide thin films by atomic force microscope nanolithography. *Applied Physics Letters*, Vol. 86, 133113–133113-2.

Jones, A. G.; Balocco, C.; King, R. & Song, A. M. (2006). Highly tunable, high-throughput nanolithography based on strained regionregular conducting polymer films. *Appl. Phys. Lett.* 89, 013119 (1-3).

Lee, H. T.; Oh, J. S.; Park, S. J.; Park, K. H.; Ha, J. S.; Yoo, H. J. & Koo, J. Y. (1997). Nanometer-scale lithography on H-passivated Si(100) with an atomic force microscope in air. *Journal of Vacuum Science Technology*, A15(3), 1451.

Lemeshko S.; Saunin S. & Shevyakov V. (2005). Atomic Force Microscope As a Tool for Nanometer Scale Surface Surface Patterning. *Nanotechnology*, 2, 719-721.

Li, S. F. Y.; Ng, H. T.; Zhang, P. C.; Ho, P. K. H.; Zhou, L.; Bao, G. W. & Chan, S. L. H. (1997). Submicrometer lithography of a silicon substrate by machining of photoresist using atomic force microscopy followed by wet chemical etching. *Nanolithography*, 8, 76-81.

Lyuksyutov, S. F.; Paramonov, P.B.; Dolog, I. & Ralich, R.M. (2003). Peculiarities of an anomalous electronic current during atomic force microscopy assisted nanolithography on n-type silicon. *Nanotechnology*, 14, 716-721 .

Lyuksyutov, S. F.; Paramonov, P. B.; Sharipov, R. A. & Sigalov, G. (2004). Induced nanoscale deformation in polymers using atomic force microscopy. *Physical Review B*, 70, 174110-174110(1-8).

Magno, R. & Bennett, B. R. (1997). Nanostructure patterns written in III –V semiconductors by an atomic force microscope. *Appl. Phys. Lett.* 70, 1855-1857.

Martin, C.; Rius, G.; Borrise, X. & Perez-Murano, F. (2005). Nanolithography on thin layers of PMMA using atomic force microscopy. *Nanotechnology*, Vol. 16, pp.1016–1022.

Notargiacomo, A.; Foglietti, V.; Cianci, E.; Capellini, G.; Adami, M.; Faraci, P.; Evangelisti, F. & Nicolini, C. (1999) Atomic Force Microscopy Lithography Study towards the Development of Nanodevices. *Nanotechnology*, 10, 458-463.

Park, J. & Lee, H.,(2004). Effect of surface functional groups on nanostructure fabrication using AFM lithography. *Material Science and Engineering* C. 24, No.1, 311-314.

Quate, C. F. (1997). Scanning probe as a lithography tool for nanostructures, *surface science*. 386, 259-264.

Sadegh Hassani, S. & Ebrahimpoor Ziaie, E. (2006). Application of Atomic Force Microscopy for the study of friction properties of surfaces. *material science: An Indian journal*, 2(4-5), 134-141.

Sadegh Hassani, S. & Sobat, Z. (2011). Studying of various nanolithography methods by using Scanning Probe Microscope. *Int .J. Nano .Dim* 1(3) , Winter 159-175.

Sadegh Hassani, S.; Sobat, Z. & Aghabozorg, H. R. (2010). Force nanolithography on various surfaces by atomic force microscope. *Int. J. Nanomanufacturing*, Vol. 5, Nos. 3/4, 217-224.

Sadegh Hassani, S.; Sobat, Z. & Aghabozorg, H. R. (2008a). Nanometer-Scale Patterning on PMMA Resist by Force Microscopy Lithography. *Iran. J. Chem. Chem. Eng.* Vol. 27, No. 4, 29-34.

Sadegh Hassani, S.; Sobat, Z. & Aghabozorg, H. R. (2008b). Scanning probe lithography as a tool for studying of various surfaces. *Nano Science and Nano Technology: An Indian journal*, Volume 2 Issue (2-3),94-98.

Samori, D. (2005). Exploring supramolecular interactions and architectures by Scanning Force Microscopies. *Chem. Soc. Rev.*, 34, 551-561.

Santinacci, L., Djenizian, T., Hildebrand, H., Ecoffey, S., Mokdad, H., Campanella, T. & Schmuki, P.,(2003). Selective Pd electrochemical deposition onto AFM-scratched silicon surfaces. Electro chimica Acta, 48, 3123-3130.

Santinacci, L.; Zhang, Y. & Schmuki, P. (2005). AFM scratching and metal deposition through insulating layers on silicon. *Surface Science*, 597(1-3), 11-19.

Sheehan, P. E. & Whitman, L. (2002). Thiol Diffusion and the Role of Humidity in Dip Pen Nanolithography. *J. Physical Review Letters*, 88, (156104-1)-(156104-4) .

Sheglov, D. V.; Latyshev, A. V. & Aseev, A. L. (2005). The deepness enhancing of an AFM-tip induced surface nanomodification. *Applied Surface Science*, 243(1-4), 138-142 .

Snow, E. S.; Campbell P.M. & Perkin F. K. (1999). High speed patterning of a metal silicide using scanned probe lithography. *Appl. Phys. Lett.* 75(101476-1478..

Sumomogi, T.; Endo, T.; Kuwahara, K. & Kaneko. R. (1995). Nanoscale layer removal of metal surfaces by scanning probe microscope scratching. *J. Vac. Sci. Technol. B*13, 1257-1260.

Sugimura, H.; Uchida, T.; Kitamura, N. & Masuhara, H.,(1993). Tip- induced anodization of titanium surfaces by scanning tunneling microscopy: a humidity effect of nanolithography. *Appl. Phys. Lett.*, 63, 1288.

Sugimura, H. & Nakagiri, N. (1997). AFM lithography in constant current mode, *Nanotechnology*, 8 A15-A18.

Sundararajan S. & Bhushan B. (1998). Micro/Nanotribological Studies of Polysilicon and SiC Films for MEMS Applications, Wear 217, 251-261.

Sundararajan, S. & Bhushan, B. (2000). Topography-Induced Contributions to Friction Forces Measured Using an Atomic Force/Friction Force Microscope. *Journal of Applied Physics*, 88(8), 4825.

Vanlandingham, M. R. (1997). The effect of instrumental uncertainties on AFM indentation measurements. *Microscopy Today* 97 (10) 12-15.

Wendel, M., Kuhn, S., Lorenz, H., Kotthaus, J. P. & Holland, M. (1994). Nanolithography with an Atomic Force Microscope for Integrated Fabrication of Quantum Electronic Devices. *Appl. Phys. Lett.* 65, 1775-1777.

Wendel, M.; Irmer, N.; Cortes, J.; Kaiser, R.; Lorenz, H.; Kotthaus, J. P. & Lorke, A. (1996), Nanolithography with an Atomic Force Microscope. *Superlattices and Microstructures* 20 (3), 349-356.

Wendel, M.; Lorenz, H. & Kotthaus. J. P. (1995). Sharpened electron beam deposited tips for high resolution atomic force microscope lithography and imagine. *Appl. Phys. Lett.* 67, 3732-3734.

Wilder, K. & Quate, C. F. (1998). Noncontact nanolithography using the atomic force microscope *Appl. Phys. Lett.*, 73 (17) 2527-2529.

Yasin, Sh.; Khalid, M. N.; Hasko, D.G. & Sarfraz, S. (2005). Correlation of surface roughness with edge roughness in PMMA resist. *Microelectronic Engineering*, 78-79, 484-489 .

Yeung, K. L. N. & Yao, J.(2004). Scanning probe microscopy in catalysis. *Nanosci. Nanotechol.*, 4, 1-44.

Yoshimura, T.; Shiraishi H.; Yamamoto J. & Okazaki Sh. (1993). Nano edge roughness in polymer resist patterns. *Appl. Phys. Lett.* 63,(6), 764-766.

9

Electrohydrodynamic Inkjet – Micro Pattern Fabrication for Printed Electronics Applications

Kyung-Hyun Choi, Khalid Rahman, Nauman Malik Muhammad,
Arshad Khan, Ki-Rin Kwon, Yang-Hoi Doh and Hyung-Chan Kim
Jeju National University
Republic of Korea

1. Introduction

In electronic industry the manufacturing of conductive patterning is necessary and ineluctable. Traditionally, lithography is widely used for fabrication of the conductive patterns. However, lithographic processes require the complicated equipments, are time consuming and the area throughput is limited. In order to reduce the material usage, process time and large area fabrication, different fabrication technique is required. Non-lithographic-direct fabrication method (Pique & Chrisey, 2001) such as inkjet (Gans et al., 2004) and roll-to-roll (Gamota et al., 2004) printing (also known as printed electronics) are predominant examples for reasonable resolution and high throughput as compared to lithography techniques. This direct fabrication technology can be further classified into two different technologies depending on the fabrication method as contact (gravure, offset or flexographic etc) and non-contact (inkjet) method. Non-contact inkjet printing method has moved beyond graphic printing as a versatile manufacturing method for functional and structural materials.

Commercially available inkjet printer can be divided into two modes based on the ejection of the fluid: Continuous, where jet emerges from the nozzle which breaks in stream of droplets or Drop-on-Demand, the droplet ejects from the nozzle orifice as required (Lee, 2002). Inkjet printing offers the advantages of low cost, large area throughput and high speed processing. The most prominent examples of inkjet printing includes the direct patterning of, printed circuit board, conductive tracks for antenna of radio frequency Identification tags (RFID) (Yang et al., 2007), Photovoltaic (Jung et al., 2010), thin film transistors (Arias et al., 2004), micro arrays of the DNA (Goldmann & Gonzalez, 2000), biosensors, etc. In case of continuous inkjet printing, the deflector directs the stream of droplets into a waste collector or onto substrate, for start and stop of the printing. This wastage of the ink issue has been addressed by the introduction drop-on-demand inkjet printing (thermal and piezoelectric). In drop-on-demand, thermal or vibration pulse are used to eject the liquid droplet from the nozzle to the substrate. However, the current printing technologies have constrained due to limitation of the ink viscosity, clogging of small size nozzles, generation of pattern smaller than the nozzle size and limitation of material to be deposited (Le, 1998). In order address these limitations, many researchers are focusing on electrohydrodynamic inkjet printing (continuous and drop-on-demand) (Park et al., 2007). Electrohydrodynamic jet printing uses electric field energy to eject the liquid from the nozzle instead of thermal or acoustic energy (Hartman, 1998). Based on the applied electric field energy, the electrohydrodynamic jetting can be used for continuous patterning, drop-on-demand printing and thin film deposition (electrospray). Electrohydrodynamic drop-on-demand, jetting or atomization has numerous applications in inkjet printing technology (Wang, 2009), thin film deposition (Jaworek, 2007), bio-application (Park 2008), mass spectrometry (Griss, 2002), etc.

2. Electrohydrodynamic jetting

In electrohydrodynamic printing the liquid is pulled out the nozzle rather than the pushing out as in case of conventional inkjet systems. When the liquid is supplied to nozzle without applying the electric field, a hemispherical meniscus is formed the nozzle due to the surface tension at the interface between the liquid and air. When the electric field is applied between the liquid and the ground plate (located under the substrate), the ions with same polarity move and accumulate at surface of the meniscus. Due to ions accumulation, the Maxwell electrical stresses are induced by the Coulombic repulsion between ions. The surface of the liquid meniscus is mainly subjected to surface tension σ_s, hydrostatic pressure σ_h and electrostatic pressure σ_e. If the liquid is considered to be a pure conductor, then the electric field will be perpendicular to the liquid surface and no tangential stress component will be acting on the liquid surface. The liquid bulk will be neutral and the free charges will confined in a very thin layer. This situation can be summarized in the following equations.

$$\sigma_h + \sigma_e + \sigma_s = 0 \tag{1}$$

Since the liquid is not a perfect conductor, the resultant electric stress on the liquid meniscus has two components, i.e. normal and tangential as shown in figure1. This repulsion force (electrostatic force) when exceeds the certain limit deforms the hemispherical meniscus to a cone. This phenomenon is known as the cone-jet transition, which refers to the shape of meniscus (Poon, 2002).

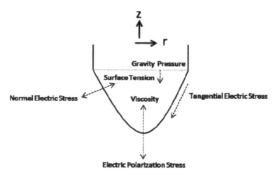

Fig. 1. Stresses due to different forces on the liquid meniscus (Kim et al. 2011)

For specific configuration and constant flow rate, there are different modes of electrohydrodynamic jetting as a function of applied voltage (Cloupeau & Foch, 1994). It should be noted that not for all liquids each mode can occur, because of the properties of the liquid. The different modes of the electrohydrodynamic jetting are discussed as follows:

2.1 Dripping, micro-dripping, spindle and intermittent cone-jet mode

In dripping mode, when the liquid is pumped in to the nozzle or capillary without applying the electric field, the droplets disintegrate from the orifice, the size of the droplets are larger than the size of the nozzle orifice. As the electric field is increased, the frequency of the droplet generation is also increased and size of the droplet decreases. At relative low flow rate, the droplet disintegrate in much smaller size as compare to the inner diameter of nozzle, this mode is known as micro-dripping mode, the frequency of the droplet increases with increase in applied electric field and size decreases. Depending on the liquid properties, increasing further electric potential, the spindle mode observed. In spindle mode, the jet extended from the meniscus and breaks up into larger droplet and satellites droplets are also observed. Further increasing in applied voltage, with relative high flow-rate intermittent cone-jet mode occurs, causing the pulsating cone-jet modes due to the high space charges reduce the electric field on the liquid jet and causing relaxation of the cone-jet into hemispherical meniscus. The pulsation in the intermittent cone-jet mode increases with increase in the applied voltage.

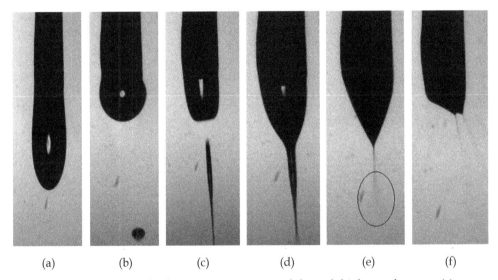

(a)	(b)	(c)	(d)	(e)	(f)

Fig. 2. Modes of electrohydrodynamic jetting captured through high speed camera (a) dripping, (b) micro-dripping, (c) spindle mode, (d) pulsating cone jet, (e) stable cone jet, and (f) multi-jet mode

2.2 Cone-jet mode

Further increase in voltage, the meniscus deforms into cone and thin stable jet emerges from the apex of the jet. This mode is known as cone-jet mode. In cone-jet mode, the intact jet used to fabricate the patterns on the surface of the substrates. The main advantage of cone-jet as compared to conventional method of ejection of the liquid is its large ratio between diameter of the nozzle and the jet. The typical jet diameters are about two orders of magnitude smaller than that of nozzle; this enables patterning at very fine resolution. However, the cone-jet also has shortcoming, it is very difficult to stabilize and control the trajectory of thin jet under an electric-field, and jet disintegrate into droplets and cause spray due to electrostatic repulsive forces between themselves.

2.3 Multi-jet mode
If the applied voltage is further increased, the cone becomes smaller and smaller. With increasing the applied voltage, second jet emerges from the cone. With further increase in applied voltage, more and more jet emerges from the cone, this mode is called multi-jet mode.

2.4 Parameters affecting the cone-jet
The parameters that influence the formation and transition of stable cone-jet mode can be divided in two groups. Operating parameters i.e. flow-rate, electric field and nozzle diameter, and liquid properties i.e. electric conductivity, viscosity and surface tension (Poon, 2002).

2.4.1 Flow-rate
The flow-rate has a significant effect on the jet diameter and stability of the jet in cone-jet transition. It is the main parameters to control the jet diameter for the patterning process. The flow-rate also affects the applied potential requirement for development of cone-jet (operating envelop) and the resulting the shape of the cone-jet. In electrohydrodynamic jetting, flow-rate also affect the stability of the jet, at low flow-rate the jet is stable, whereas the high flow-rate in cone-jet region destabilize the jet resulting in shorter jet length. This is due to amount of charge carrying at high flow-rate which destabilizes the jet. The typical effects of the flow-rate on stable cone-jet mode are shown in figure 3.

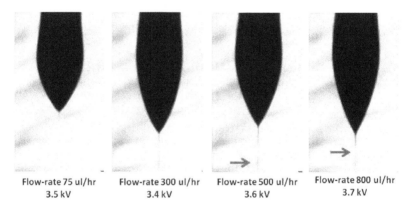

| Flow-rate 75 ul/hr | Flow-rate 300 ul/hr | Flow-rate 500 ul/hr | Flow-rate 800 ul/hr |
| 3.5 kV | 3.4 kV | 3.6 kV | 3.7 kV |

Fig. 3. Effect of flow rate on the cone-jet transition, as shown the jet diameter increases with increase in flow-rate. The red arrow indicates the jet break-up point.

2.4.2 Electric field (applied voltage)
Electric field affects the morphology of the cone, by increasing the electric field strength in steady cone-jet mode, the cone-jet recedes towards the nozzle. However, there is less effect on the jet diameter by increasing the applied voltage. At relatively low flow-rate and high electric field, the jet disintegrated into mist of small droplets also known as electrospray

atomization, this behavior is used for the thin film deposition of functional material. The typical shape of the come at different applied voltage is shown in figure 4.

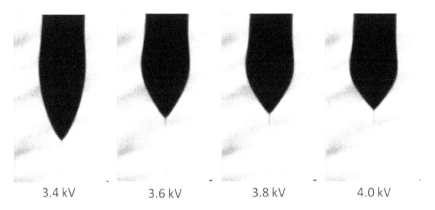

| 3.4 kV | 3.6 kV | 3.8 kV | 4.0 kV |

Fig. 4. Effect of the applied voltage on shape of the cone-jet at constant flow rate (200µl/hr)

2.4.3 Nozzle diameter
Nozzle diameter has significant influence on the operating envelope of the stable cone-jet region. Smaller nozzle diameters extend the lower and higher value of flow-rate limit. The voltage requirement for the cone-jet also decreases with decrease in nozzle diameter for any given liquid.

2.4.4 Conductivity
The liquid conductivity affects both the shape of the liquid cone and stability of the jet, due to the amount of electric charge on the liquid surface and jet produces is also very unstable because of high radial electric field. Highly conductive liquid deforms into sharp cone-jet shape. Liquid with very low conductivity do not deform into cone-jet by applying electric field, only dripping mode is observed. Liquids with intermediate conductivity range produce steady cone-jet.

2.4.5 Viscosity
The role viscosity is in the stabilization of the jet and diameter of the jet produced. In high viscous liquid, the jet is stable for larger portion of the length but also produces the thicker diameter. This is due to charge mobility, which is reduced significantly in high viscosity liquid, and causes decrease in conductivity.

2.4.6 Surface tension
The formation of the jet occurs when the electrical forces overcomes the surface tension on the apex of the meniscus. The required applied voltage will be increased with increase in surface tension of the liquid.

2.5 Operating envelop
In order to perform the patterning of any liquid containing nanoparticles, the operating parameters of flow rate corresponding to applied voltage for stable cone-jet has to be

determined. Starting with high flow rate formation of stable cone-jet is determined by applying different voltages. Then for each flow-rate, the range of applied voltage is investigated at which the stable cone-jet is observed. This creates an operating envelop of certain liquid in electric field and voltage domain. Based on the operating envelop, behavior of the jetting is observed and parameters for patterning are determined. Operating envelop along with different modes of electrohydrodynamic jetting of the liquid containing Copper nanoparticles is shown in figure 5.

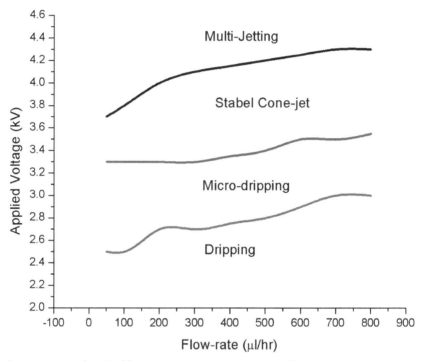

Fig. 5. Operating envelop (stable con-jet region) along with different electrohydrodynamic jetting mode of ink containing copper nanoparticles, using metallic capillary of inner diameter 410μm, external diameter 720μm and capillary to ground distance is 2mm.

3. Patterning setup

For direct patterning through electrohydrodynamic printing, lab developed system was used. The equipment used for patterning is consistent of high voltage power supply, function generator, 5 channel voltage distributor, syringe-pump for ink supply, X-Y stage with motor controller, substrate holder, high-speed camera and nozzle holder with Z-axis controller, which are connected to National Instruments PXI-1042Q hardware system. Which is controlled through lab made software based on LabView. Positive potential is applied to the nozzle head while ground is applied to the conductive plate. The substrate is place on the top of metallic ground plate. The schematic of experiment setup along with the actual system is shown in Figure 6.

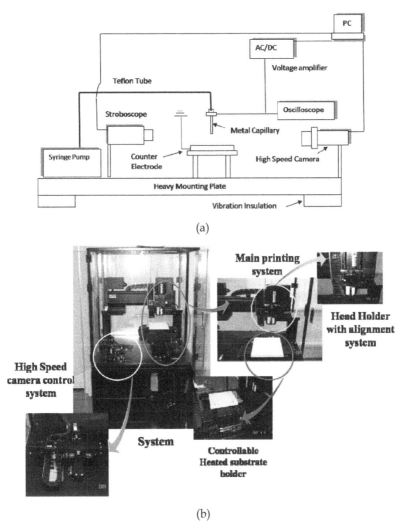

Fig. 6. (a) Schematic of lab developed system and (b) Photograph of the electrohydrodynamic inkjet system used for patterning

4. Electrohydrodynamic printing

Electrohydrodynamic printing can be divided in two different categories, depending on the ejection of liquid from the nozzle, continuous and drop-on-demand. The experiment setup for both the modes is same.

4.1 Continuous electrohydrodynamic printing

For continuous printing, ink containing copper nanoparticles is used. Initial patterning is performed by using metallic capillary of internal diameter of 210μm and outer diameter of

410µm. The distance between capillary and ground is kept 1.5mm. Patterning is performed on the glass substrate of 0.5mm thickness placed on the top of metallic ground plate by applying the lower limit of the applied voltage (ranging from 3.1kV to 3.6kV) and corresponding applied flow-rate, which is investigated for stable cone-jet operating envelop, the speed of substrate is kept at 25mm/sec for all the experiment. In electrohydrodynamic printing the jet diameter is more dependent on the flow-rate, in a cone-jet region the diameter of the jet increases with increase in flow-rate. The applied voltage has minor effect on the diameter of the jet in cone-jet mode. After patterning, the samples are placed in dry oven for sintering at 80ºC for 30min. After sintering the pattern width is measured with the help of digital microscope. The pattern width with respect to applied flow-rate is shown in graph at figure 7. As shown in the graph the pattern width increases with increase in the applied flow-rate. However, pattern at high flow-rate is irregular as compared to pattern at low flow-rate, because the length of the jet decreases with increase in flow-rate at cause destabilization to the jet in cone-jet mode. The minimum pattern width of 116µm is achieved after the sintering.

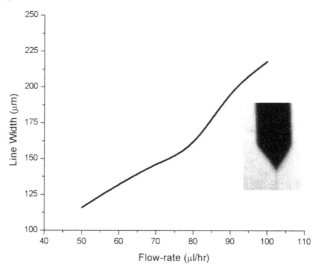

Fig. 7. Pattern width with respect to flow-rate on glass substrate

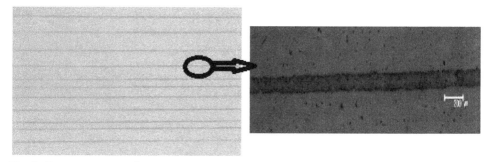

Fig. 8. Camera image and microscopic image of the copper pattern on glass substrate

Since the substrate i.e. glass has a hydrophilic surface; as a result the deposited jets were able to spread out so that the width of the lines became larger than the original size of the generated jets. Figure 8 shows the high zoom static camera and optical microscope image of continuous copper tracks on glass substrate without any defects such as bulges or coffee-ring effects.

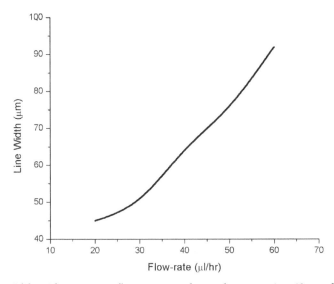

Fig. 9. Pattern width with respect to flow-rate on glass substrate using 60μm glass capillary

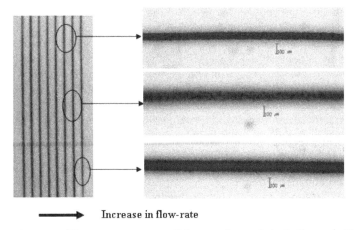

Increase in flow-rate

Fig. 10. Pattern images of the copper nanoparticles on glass substrate through 60μm internal diameter glass capillary nozzle

For application in printed electronics, higher resolution pattern size is required. In order to reduce the pattern size, the pattern is performed through tapered glass capillary of 60μm inner diameter and 80μm is used. The glass capillary tubes of 1.5mm outer diameter and

0.75mm inner diameter (BF 150, Sutter Instrument) was pulled and micro-nozzles is formed, with sharp tip by using a micropipette puller (P-97, Sutter Instrument). The voltage is applied by inserting the copper wire of 500µm in the glass capillary and connected to high voltage power supply. Before performing the patterning, the operating envelop for stable cone-jet is investigated. The gap between nozzle and glass substrate is also reduced to 500µm, because of small size of jet produces and is difficult to control the smaller diameter jet at larger length. The applied voltage range for developing the stable cone-jet is also reduces (ranging from 1.8kV to 2.5kV) with respect to applied flow-rate (ranging from 20µl/hr to 80µl/hr). The pattern width with respect to flow-rate by applying minimum value of required voltage, after sintering is shown in figure 9, the minimum pattern width achieved is approximately 42µm.

Figure 10 shows the image of the pattern on the glass substrate along with the microscopic images of the pattern after sintering. The images show that the pattern width of copper nanoparticles increases with increase in flow-rate.

4.2 Drop-on-demand electrohydrodynamic printing

In continuous electrohydrodynamic printing mode, the stabilization of the micron size jet is very difficult (Hohman, et al., 2001), and also problems in the placement of the jet at start point and end point of the pattern. In order to address these issues, the electrohydrodynamic drop-on-demand printing is the alternative technology. In electrohydrodynamic drop-on-demand mode, the printing is performed through time dependent generation of cone-jet by applying the pulsed voltage to the liquid in capillary or nozzle. When the pulse voltage is applied to the capillary, the meniscus of the liquid deforms into cone-jet and generating a thin jet, as the voltage is switched-off the jet breaks up and generates small droplet, and this phenomena is pulsating by generating single droplet at each pulse. The size of droplet and the frequency of the droplet depend on the amount of pulsed voltage applied, frequency of the pulse, diameter of the nozzle, viscosity of the liquid and conductivity of the liquid (Stachewicz et al., 2009). In previous researches, for drop-on-demand, the researchers have applied simple square wave pulse voltage or by superimposing AC on applied DC voltage, in both the cases the pulse is square either with zero or bias-voltage (Li et al., 2009; Kim et al., 2008). This square pulse induces unnecessary vibrations in meniscus causing problems in the placement of the droplet on the substrate. In order to avoid this vibration in meniscus, multi-step voltage is suggested for electrohydrodynamic drop-on-demand printing (Rahman et al., 2011; Kim et al. 2011). Multi-step voltage is applied by super-imposing two square waves with same frequency but different duty cycle on each other. The multi-step voltage is consist of bias-voltage "V_a" for initialization of the meniscus; intermediate-voltage "V_b" for deformation of the meniscus into cone shape and ejection-voltage "V_c" for steady droplet generations. The applied voltage is in the form of two step functions, with "V_a" consists of 25% of the pulse, "V_b" consists 50% of the pulse and "V_c" consists 25% of the pulse.

The ejection behavior at 75µl/hr at 200Hz frequency by applying square voltage (V_a=2kV and V_b=3kV) with 50% of Duty Cycle and multi-step voltage (V_a=2kV, V_b=2.5kV and V_c=3kV) is shown in figure 11 through pictures taken with high speed camera. The main benefit of the multi-step pulse voltage as compared to the square pulse voltage is time of application of bias voltage to ejection voltage. As shown in figure 11, in square voltage there is sudden change in applied voltage to ejection voltage, which causes disturbance in meniscus and instability in ejection phenomena. By multi-step pulse voltage there is intermediate voltage which ramp the effect of applied voltage, which avoids the sudden application of high voltage to the meniscus and also induced less vibration to the meniscus hence stabilization of the ejection process. The other advantage of the multi-step pulse voltage is the related to high voltage switching hardware, in square the switching time is less as compared to multi-step pulse voltage due to intermediate voltage.

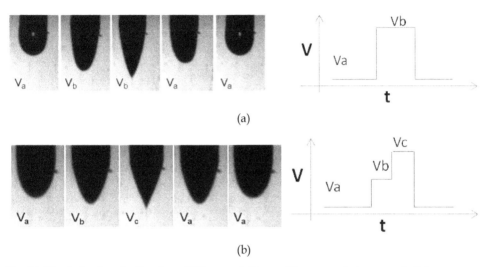

(a)

(b)

Fig. 11. The behavior ejection from 210µm metallic capillary, nozzle to ground distance 1.5mm at flow-rate 75µl/hr and frequency 200Hz (a) Square wave V_a=2kV and V_b=3kV with 50% Duty Cycle and (b) Multi-step voltage V_a=2kV, V_b=2.5kV and V_c=3kV (Rahman et al. 2011)

For drop-on-demand patterning on the glass substrate, initial experiments were performed to find lower and upper values for the applied DC voltage at which stable cone-jet formed and corresponding flow-rate. The average lower value of DC voltage at which cone-jet formed was 3.2kV and average upper value is 4.1kV. The dimension of the pattern was measured through digital microscope after sintering the Ag nanoparticles ink for 1hour at 250°C. Following can be described as input parameters for drop-on-demand printing:

1. "V_a" bias-voltage must be closed to the value of lower limit of stable cone-jet region.
2. "V_b" intermediate-voltage must be at middle range of stable cone-jet region.
3. "V_c" ejection-voltage must be closed to the value of upper limit of stable cone-jet region.
4. Determine the optimal value of "V_a", "V_b" and "V_c", at low flow-rate and frequency at which the phenomena is stable.
5. After investigating the optimal value of "V_a", "V_b" and "V_c", optimal frequency and flow-rate should be obtained.
6. Determine the optimal substrate speed based on the drop-on-demand frequency and droplet spacing on the substrate.

Figure 12 shows the static high zoom camera and microscopic images of printed dots and lines on the glass substrate by using EHD drop-on-demand ink jet printing technique of ink containing copper nanoparticles, using metallic capillary of internal diameter 410µm and external diameter 720µm. Patterning has been carried out at a constant flow rate and at pulse frequencies of 10Hz, 25Hz, 50Hz, and 100Hz with a constant linear motor speed (substrate speed) of 25mm/s. Effect of biased and pulse voltages on droplet size has been analyzed by varying the biased and step voltages. figure 12(a) shows the deposited droplets which are generated at applied pulse of 50Hz frequency with 50% duty cycle and at voltages V_a, V_b and V_c of 2.5kV, 4.5kV (first pulse of 2kV) and 5.5kV (second pulse of 1.5kV) respectively. The average diameter size of droplets is 780µm. Comparatively smaller droplets have been generated by increasing the magnitude of biased voltage and decreasing the magnitude of step voltages. The deposited droplets in figure 12(b) are generated at the same frequency and duty cycle as that of figure 12(a) i.e. 25Hz and 50% respectively but at low pulse voltages i.e. V_a, V_b and V_c; 3.5kV, 4.5kV (first pulse of 1kV) and 5.0kV (second pulse of 0.5kV) respectively. The average diameter size of deposited droplets is 780µm. The reason of this relatively smaller drop generation is that with low biased voltage and high pulsed voltages, energy gain per unit area of the liquid and the tangential electric stress at the liquid meniscus increases more quickly than the normal electric stress. As a result, greater pulsed voltages (V_b and V_c) are more likely to produce a temporary jet rather than a drop-on-demand mode which generates relatively large sized droplets. Similarly figure 12(c) shows the patterned droplets which are generated at applied pulse of 10 Hz frequency with 75% duty cycle and at voltages V_a, V_b and V_c 3.5kV, 4.5kV(first pulse of 1kV) and 5.0kV(second pulse of 0.5kV) respectively. The deposited droplets have an elliptical shape rather than round due to high duty cycle of the pulse voltage. High duty cycle (75%) increases the application time of the triggering pulse which results in a temporary jet rather than a droplet. Since the substrate speed is constant and relatively high than the speed of ejection of temporary jet which does not allowing the temporary jet to accumulate to a large round shape drop on the substrate. As a result, temporary generated jet forms an oval shape drop after deposition on the substrate. Using this high duty cycle (75%) of pulsed voltage, conductive lines are patterned at applied pulse of 100 Hz frequency and at voltages V_a, V_b and V_c of 3.5kV, 4.5kV (first pulse of 1kV) and 5kV (second pulse of 0.5kV) respectively. The printed lines shown in figure 12(d) have an average size of 780µm. It can be concluded from the printed results shown in figure 12(d) that EHD drop-on-demand can also be used for printing of conductive lines for metallization in printed circuit boards and backplanes of printable transistors if the substrate speed and frequency of droplet generation get synchronized.

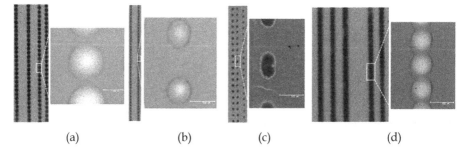

(a)   (b)   (c)   (d)

Fig. 12. Camera and microscopic images of deposited droplets and line patterning resulted by drop-on-demand ejection at constant flow rate and constant substrate speed: (a) 50Hz and 50% duty cycle (b) 25Hz and 50% duty cycle (c) 10Hz and 75% duty cycle (d) the line pattern by drop-on-demand at 100Hz and 75% duty cycle (Kim et al. 2011)

The drop-on-demand experiment is also performed by using silver nanoparticles ink. The numbers of experiments are performed by changing the value of V_a, V_b and V_c. The frequency of the multi-step voltage is at 50Hz and with flow-rate 75µl/hr. The droplet diameter on the substrate at different multi-step voltage is shown in table 1. As shown in table 1, in Case-1 the droplet diameter is larger due to spray because at high voltage the droplet caries more charge. As in Case-5 and 6 the droplet diameter is smaller due to more stable phenomena. The microscopic image of the printed droplet on glass substrate after sintering for Case-1 and Case-6 is shown in figure 13.

Case	V_a (kV)	V_b (kV)	V_c (kV)	Approx. Droplet Diameter (µm)
1	0	2	4.1	200
2	1	2	4.1	189
3	2	3	4	160
4	3	3.5	4	126
5	3	3.2	3.7	106
6	3.2	3.5	3.8	82

Table 1. Droplet Diameter at different applied multi-step voltage

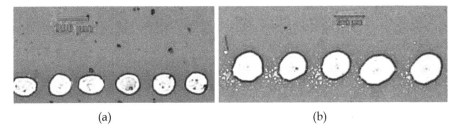

(a)   (b)

Fig. 13. (a) case-1 and droplet diameter approximately 200µm and   (b) case-6 and droplet diameter approximately 82µm (Rahman et al. 2011)

In order to analyze the effect of frequency on the drop-on-demand pattering, the experiments are performed by changing the applied frequency. The applied multi-step voltage is kept as Case-6 (V_a=3.2kV, V_b= 3.5kV and V_c=3.8kV) and flow-rate 75µl/hr. The measured droplet diameter on the glass substrate against the frequency is shown in graph at figure 14. The droplet diameter is decreased from 120µm to 40µm as the frequency increases from 10Hz to 350Hz. The maximum applied frequency at which the drop-on-demand phenomena observed is 350Hz for the silver ink by using 210µm inner diameter capillary. The reason is due to short pulse times at high frequency, the voltage required to generate the jet is applied in shorter time i.e. jetting time decreases due to which droplet diameter decreases at higher frequencies. Figure 15 shows the microscopic images of the droplets after sintering by applying 10Hz and 350Hz frequency. The result also indicates the size of the droplets is much smaller than the nozzle size which is the main advantage of electrohydrodynamic drop-on-demand technique.

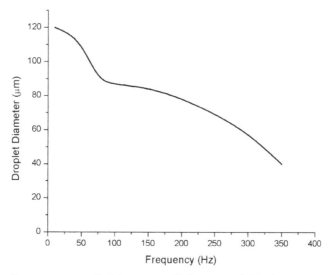

Fig. 14. Droplet diameter vs. applied frequency (Rahman et al. 2011)

(a)   (b)

Fig. 15. Microscopic images of the sintered droplet by applying V_a=3.2kV, V_b=3.5kV, V_c=3.8kV and flow-rate 75µl/hr (a) applied frequency 10Hz and diameter approximately 120µm (b) applied frequency 350Hz and droplet diameter approximately 40µm (Rahman et al. 2011)

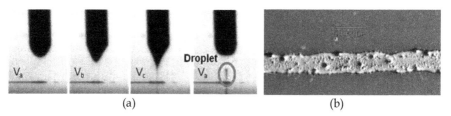

(a)   (b)

Fig. 16. (a) Sequential images of the drop-on-demand pattern on the glass substrate with respect to applied multi-step pulsed voltage and (b) microscopic image of the line pattern after sintering process (Rahman et al. 2011)

The pattern line is formed after synchronizing the substrate speed with the drop-on-demand frequency by analyzing the droplet spacing on the substrate. The sequential images of the drop-on-demand line patterning with respect to applied multi-step pulse voltage at V_a=3.2kV, V_b=3.5kV, V_c=3.8kV, flow-rate 75µl/hr and frequency 100Hz along with the pattern line are shown in figure 16. The pattern size shown is approximately 95μm, the line pattern size is greater than the droplet size (87μm), which is due to the over lapping of the droplets to create the line pattern.

Figure 17 shows the XRD spectrum of the printed line pattern on the glass substrate. The XRD spectrum peaks shows the existence of the silver only which confirms the deposited material was consist of silver nanoparticles.

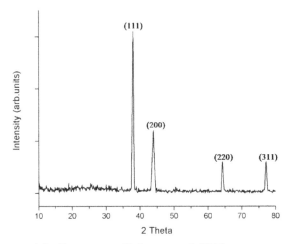

Fig. 17. XRD spectrum of the line pattern (Rahman et al. 2011)

The resistance of the pattern line is measured by 4-point probe method by measuring the voltage drop ΔV across 2mm long segment of the line pattern by applying the different current intensity of 10μA, 20μA, 50μA, 75μA and 100μA. The pattern line showed the linear ohmic behavior with resistance of 0.39Ω. I-V curve obtain through 4-point measurement is shown in figure 18.

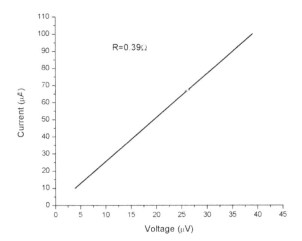

Fig. 18. I-V curve of the pattern line (Rahman et al. 2011)

4.3 Multi-nozzle electrohydrodynamic printing

The main draw-back of single nozzle electrohydrodynamic printing is the limitation of low throughput. In order to address this drawback and attain high production efficiency electrohydrodynamic inkjet printing process for industrial production of printed displays, PCBs, printed TFTs and printed solar cells, many researcher are working on a multi-nozzle electrohydrodynamic inkjet printing process(Lee et al., 2008). However, due to the interaction (cross-talk) between the electrically charged neighboring jets, it is difficult to precisely control and reproducible multi-nozzle EHD inkjet printing process. To overcome the limitation of low throughput of EHD inkjet printing process, a multi-nozzle EHD inkjet printing head consisted of three nozzles is fabricated and successfully tested by printing simultaneously conductive lines of silver nanoparticles ink onto glass substrate (Arshad et al 2011). Multi-nozzle electrohydrodynamic inkjet printing is consisted of three parts i.e. PDMS (Polydimethylsiloxane) holder, glass capillaries and copper electrodes. PDMS holder is manufactured through molding technique with the channels for capillaries, ink-supply and electrodes. The schematic of the mold and multi-nozzle head is shown in figure 19(a) and (b). The tapered glass capillaries of 100µm internal diameter and 120µm are then inserted in the outlet channels of the PDMS part. Finally, three copper electrodes having outer diameter of 500µm are inserted from the top of the PDMS holder. The nozzle to nozzle distance is kept at 3mm. The complete schematic of the multi-nozzle EHD inkjet printing head is shown in Figure 19(c).

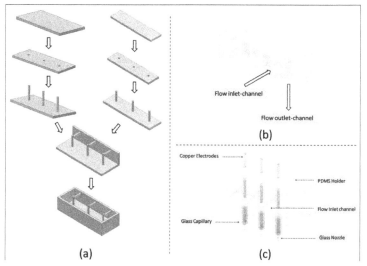

Fig. 19. Schematic illustration of the multi-nozzle EHD inkjet printing head fabrication process: (a) Simplified fabrication steps of mold preparation (b) Resulted PDMS holder having L-shaped channels for ink supply and clasping of glass capillaries (c) Complete multi-nozzle EHD inkjet printing head (Khan et al 2011)

Figure 20 shows images of stable meniscus formed at the tip of individual nozzles. Cone-jets are formed at the tip of each nozzle i.e. nozzle 1, nozzle 2 and nozzle 3 at an applied DC voltage and flow rate of 3.5kV and 20µl/h respectively.

Fig. 20. Photographs of axisymmetric cone-jet established at the nozzles tips of multi-nozzle electrohydrodynamic inkjet printing head (Khan et al 2011)

Printing is performed by applying a DC voltage of 3.5kV and flow rate of 20µl/h to each nozzle. The nozzles to substrate distance is set as 300µm while substrate speed is kept constant at 10mm/sec. Figure 21 shows the high zoom static camera and optical microscope image of continuous silver tracks simultaneously printed by three nozzles on glass substrate without any defects such as bulges or coffee-ring effects. The average line width of the printed lines is 140µm.

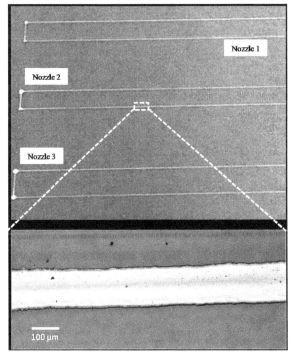

Fig. 21. Camera and optical microscope image of continuous silver patterns printed on glass substrate by multi-nozzle printing at an applied voltage of 3.5 kV and flow rate of 20µlh (Khan et al 2011)

Moreover, the SEM images of a typical printed line as shown in figure 22 also illustrates that nanoparticles are three-dimensionally interconnected with each other, which favorably affect the electrical conductivity.

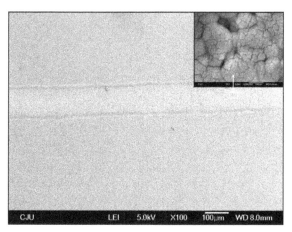

Fig. 22. SEM image of continuous silver pattern deposited by multi-nozzle electrohydrodynamic inkjet printing (Khan et al 2011)

4.4 Thin film deposition

The main benefit of the electrohydrodynamic atomization in cone-jet mode also known as electrospray deposition can be used for thin film deposition for functional materials on different substrates. The same experiment setup for printing is applicable for the thin film deposition through electrospray. In electrospray mode, the nozzle to substrate distance is larger than the patterning setup; this allows the jet to disintegrate into small monodisperse droplets of equal size due to the repulsive forces in the droplets carrying the charges. These mondisperse droplets landed on the substrate and generate the uniform thin layer of that material. The thickness and area of the layer depends on the flow-rate, distance between nozzle and substrate, and also the time of the spray.

For thin film deposition of CIS (Copper-Indium di-Selenide) through electrospray deposition is performed by using 430µm internal diameter metallic nozzle (Muhammad et al., 2010). Initially the operating envelop for stable cone-jet is investigated for the ink containing CIS nanoparticles. The operating envelop for CIS ink along with different electrohydrodynamic modes is shown in figure 23. For spray purpose higher value of applied voltage is selected. Different experiments are performed by changing the flow rate and also the distance between nozzle and substrate distance. The quality and thickness of deposition of thin layers using ESD depends upon three main factors i.e. distance from nozzle to substrate, spraying time or substrate speed and the flow rate. Therefore as less is the distance between the nozzle and the substrate, higher will be the layer quality and also the layer thickness will be on the higher side and vice versa. Similarly as much is the spraying time, or as less is the substrate speed in this case, the layer quality will be on the higher side, however concerns will be on the thickness of the layer which can negatively affect the device efficiency. Figures 24 present the layer morphology obtained by SEM at a flow rate of 150µl/h and varying nozzle to substrate distances with substrate speed of

0.5mm/s. As clear from the figures, the best layer quality is achieved at a nozzle to substrate distances of 6 mm and 8 mm the dense layers is produced and particles are completely intact and no pores or islands are visible as in the case of nozzle to substrate distances of 13mm the layer is porous and contain voids. These pores will case defect in the functionality of the deposited layer.

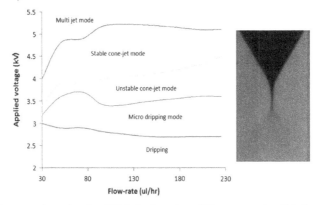

Fig. 23. Operating envelope for the CIS ink, showing different modes (Muhammad et al., 2010)

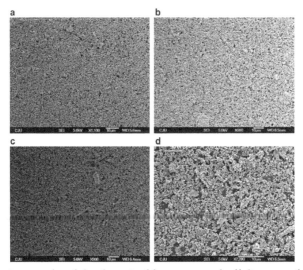

Fig. 24. FE-SEM micrographs of the deposited layer at stand-off distance of a. 6 mm, b. 8 mm, c. 10 mm, d. 13 mm, at substrate speed of 0.5 mm/s and flow rate of 150 ml/h (Muhammad et al., 2010)

5. Conclusions

Electrohydrodynamic inkjet printing is relatively new but very power tool and process for the direct patterning of the functional materials on substrate. Electrohydrodynamic inkjet printing can be used in continuous mode as well as drop-on-demand mode. The main benefit of the electrohydrodynamic inkjet printing is generation of pattern smaller than the nozzle size, because the printing is performed by pulling the liquid rather than pushing of the liquid, which is limitation in other inkjet technology such as thermal and piezoelectric printing. The advantage of electrohydrodynamic printing over conventional printing is high resolution can be obtained. The other advantage of this direct patterning method is the flexibility in the process in terms of material to be used as well as patterns can be made on different kinds of substrate. In continuous mode the patterning is performed by achieving through thing continuous jet in the stable cone-jet mode, however the thin is difficult to stabilize due to electric field. The drop-on-demand mode can be used as alternative to continuous electrohydrodynamic printing. In drop-on-demand the stable cone-jet is achieved for shorter period of time by applying the pulse voltage. Single nozzle electrohydrodynamic printing has low throughput, in order to increase the efficiency multi-nozzle printing can be performed as reported. The other advantage of electrohydrodynamic printing is thin film deposition in electrospray mode. Thin films can be deposited on the surface of different substrate without changing the experimental setup. This combination of both the technologies (patterning and thin film deposition) can help in fabrication of the electronic devices such as TFT, OLED or Solar Cells etc. through single technology.

6. Challenges and future trends

There is lot of research is being performed in the field of the electrohydrodynamic printing by patterning the functional materials even in submicron level (Schirmer et al. 2010). However the functional materials meeting these requirements are a challenge, which can be used for the patterning purpose. The inks containing these functional materials impact the morphology, adhesion, chemical and environmental stability, these factors affect the performance. In addition to this the material used in direct patterning technologies has drawbacks in functional performance such as ion mobility, on-off voltage, threshold voltage and off current. Many researchers are working in the field of chemical and material technologies to overcome these limitations, by introducing the different materials that can be used for fabrication of devices through direct fabrication technology. Inorganic material such as carbon nanotube (single-wall and multi-wall) and metal nanoparticles (silver, gold and copper) based inks and pastes are commercially available for the direct patterning applications. Recently there researches are also working on the organic materials that can be used as the insulation material for electronic devices and also conductive organic material such as PEDOT, which can used for the fabrication of the conductive tracks. Recently composite polymers have received interests in the field of direct patterning material due to improvement in mechanical, thermal, optical and conductive properties. But the major drawback is still the performance of devices fabricated through the direct patterning technology as compared to conventional electronic.

In electrohydrodynamic inkjet printing, for large area manufacturing the design of multi-nozzle printing head is another bottle neck due to interfering of electric field between nozzles. The miniaturization of the nozzles dimensions cause problems in obtaining the symmetric and stable cone-jet for printing. Other limitation is the fabrication of such small size nozzles. To optimize the performance of small nozzles, the physics and chemistry of the nozzle has to optimize to ensure the printability of different functional materials, with damaging the nozzles. When using the micron size nozzle, the issue of the clogging arises due to accumulation of the nanoparticles on the nozzle opening and cause blocking, which affects the performance of the printing. For micron size nozzle special inks are required from which printing can be performed.

In general, electrohydrodynamic inkjet printing is powerful tools for direct patterning of the functional materials and can be used for high resolution printing. Developing this technology will allow exploring the potential application of electrohydrodynamic printing in high resolution fabrication of electronic devices and appear to be promising direction for the future.

7. Acknowledgement

This work was supported by the National Research Foundation of Korea (NRF) grant funded by the Korea government (MEST) (No.2010-0026163).

8. References

Pique, Alberto. & Chrisey, D. B. (2001). *Direct-Write Technologies for Rapid Prototyping: Application to Sensors, Electronics, and Passivation Coatings*, ISBN 978-0121742317, San Diego, USA

Gamota D., Brazis. P., Kalyansundaram, K. & Zhang, J. (2004). *Printed Organic and Molecular Electronics*. ISBN 1-4020-7707-6, Massachusetts, USA

Gans, B. J., Duineveld, P. C. & Schubert, U. S. (2004). *Inkjet printing of polymers: State of the art and future development*. Advanced Materials, 16, 203–213.

Lee, E., R. (2003). *Microdrop Generation*. ISBN 0-8493-1559-X, Florida USA

Yang, L., Rida, A., Vyas, R. & Tentzeris,M. M. (2007). *RFID Tag and RF Structures on a Paper substrate Using Inkjet-Printing Technology*. IEEE Transactions on Microwave Theory and techniques, vol. 55, no. 12

Jung, J., Kim, D., Lim, J., Lee, C. & Yoon, S. C. (2010). *Highly Efficient Inkjet-Printed Organic Photovoltaic Cells*. Japanese Journal of Applied Physics, 49, 5, 05EB03-05EB03-5

Arias, A. C., Ready, S. E., Lujan, R., Wong, W. S., Paul, K. E., Salleo, A., Chabinyc, M. L., Apte, R. & Street R. A. (2004). *All jet-Printed Polymer Thin-Film Transistor Active-Matrix Backplanes*. Applied Physics Letters, 85, 15, 3304-3306.

Goldmann, T., Gonzalez, J. S. (2000), DNA-Printing: *Utilization of a Standard Inkjet Printer for the Transfer of Nucleic Acids to Solid Supports*. Journal of Biochemical and Biophysical Methods, 42, 105-110.

Le, H. P. (1998). *Progress and Trends in Ink-jet Printing Technology*. The Journal of Imaging Science and Technology, 42, 1, 49-62

Park, J.U., Hardy, M., Kang, S.U. , Barton, K., Adair, K., Mukhopadhyay, D.K., Lee, C.Y., Strano, M. S., Alleyne A. G., Georgiadis, J. G., Ferreira, P. M. & Rogers, J. A. (2007). *High-Resolution Electrohydrodynamic Jet Printing*. Nature Materials, 6, 782-789

Hartman, R. P. A. (1998). *Electrohydrodynamic Atomization in the Cone-jet Mode from Physical Modeling to Powder Production*. PhD Thesis, TU Delft

Wang, K., Paine, M. D. & Stark, J. P. W. (2009). *Fully Voltage-Controlled Electrohydrodynamic Jet Printing of Conductive Silver Tracks with a Sub-100μm Linewidth*. Journal of Applied Physics, 106, 024907

Jaworek, A., (2007). *Micro- and Nanoparticle Production by Electrospraying*. Powder Technology, 176, 18–35

Park, J.U., Lee, H. L., Paik, U., Lu, L., & Rogers, J. A. (2008). *Nanoscale Patterns of Oligonucleotides Formed by Electrohydrodynamic Jet Printing with Applications in Biosensing and Nanomaterials Assembly*. Nano Letters, 8 (12), 4210-4216

Griss, P., Melin, J., Sjodahl, J., Roeraade, J., Stemme, G., (2002). *Development of Micromachine Hollow Tips for Protein Analysis Based on Nanoelectrospray Ionization Mass Spectrometry*. Journal Micromechanics and Microengineering, 12, 682–687

Poon, H. F. (2202). *Electrohydrodynamic Printing*. PhD thesis, Princeton University

Cloupeau, M. & Foch B. P., (1994). *Electrohydrodynamic Spraying Functioning Modes: A Critical Review*. Journal of Aerosol Science, 24 (6), 1021-1036

Hohman, M. M., Shin, M. Rutledge, G. & Brenner, M. P. (2001). *Electrospinning and electrically forced jets. I. Stability theory*. Physics of Fluid, 13 (8), 2201-2220

Chen, C. H., Kelder, E. M., Van-Der-Put J.J.M. & Schoonman, J. (1996). *Morphology Control of Thin LiCoO$_2$ Films Fabricated Using the Electrostatic Spray Deposition (ESD) technique*. Journal of Materials Chemistry, 6, 765-771

Stachewicz U., Yurteri C.U., Marijnissen J.C.M., & Dijksman, J.F., (2009). *Stability Regime of Pulse Frequency for Single Event Electrospraying*. Applied Physics Letters 95, 224105

Li, J.L., & Zhang, P., (2009). *Formation and Droplet Size of EHD Drippings Induced by Superimposing an Electric Pulse to Background Voltage*. Journal of Electrostatic, 67, 562-567

Kim, J. H., Oh, H. C., & Kim, S. S. (2008). *Electrohydrodynamic Drop-on-Demand Patterning in Pulsed Cone-Jet Mode at Various Frequencies*. Journal of Aerosol Science, 39, 819-825

Rahman, K., Khan, A., Nam, N. M., Choi, K. H. & Kim D. S. (2011). *Study of Drop-on-Demand Printing Through Multi-step Pulse Voltage*. International Journal of Precision Engineering and Manufacturing, 12 (4)

Kim D. S., Khan, A., Rahman,, K., Khan, S., Kim, H. C. & Choi, K. H. (2010). *Drop-on-Demand Direct Printing of Colloidal Copper Nanoparticles by Electrohydrodynamic Atomization*. Materials and Manufacturing Processes (doi: 10.1080/10426914.2011.551956)

Lee J.S., Kim S.Y., Kim Y.J., Park J., Kim Y., Hawng. J., & Kim Y.J. (2008). *Design and Evaluation of a Silicon Based Multi-Nozzle for Addressable Jetting Using a Controlled Flow Rate in Electrohydrodynamic Jet Printing*. Applied Physics Letter, 93, 243114

Khan A., Rahman K., Hyun M.T., Kim D.S., & Choi K.H. (2011). *Multi-Nozzle Electrohydrodynamic Inkjet Printing of Silver Colloidal Solution for the Fabrication of Electrically Functional Microstructure*. Applied Physics A (doi:10.1007/s00339-011-6386-0)

Muhammad N. M., Sundharam, S., Dang H. W., Lee, A., Ryud B. H. & Choi K. H. (2010). *CIS Layer deposition through electrospray process for solar cell fabrication*. Current Applied Physics, 11 (1), S68-S75

Niklas C. Schirmer, N. C., Kullmann, C., Schmid, M. S., Burg, B. R., Schwamb, T., & Poulikakos D. (2010). *On Ejecting Colloids Against Capillarity from Submicrometer Openings: On-Demand Dielectrophoretic Nanoprinting*. Advanced Materials, 22, 4701-4705

10

A Feasible Routine for Large-Scale Nanopatterning via Nanosphere Lithography

Zhenyang Zhong, Tong Zhou, Yiwei Sun and Jie Lin
Fudan University/Department of Physics Shanghai,
China

1. Introduction

Due to the unique electronic, optical, catalytic and biological properties, well ordered nanostructures have attracted enormous interest. They have potential applications in photonic crystal devices (Yablonovitch, 1987), large-density magnetic recording devices (Chou et al., 1994), novel electronic devices (Schmidt & Eberl, 2001), synthesis of DNA electrophoresis mediate (Volkmuth & Austin, 1992) · nanocontainers (Chen et al., 2008), surface-plasmon resonance biosensors (Brolo et al., 2004), antireflective coatings for solar cells (Yae et al., 2005), and etc. Such broad applications of nanostructures were intimately associated with their unique properties, which are sensitively dependent on their size and/or shape. It is well-established that magnetic (Shi et al., 1996 ; Zhu et al.,2004), optical (Aizpurua et al., 2003; Larsson et al. 2007), electrocatalytic (Bratlie et al., 2007 ; Narayanan & El-Sayed, 2004), optoelectronic (Chovin et al., 2004), data storage (Ma, 2008), thermodynamic (Volokitin et al., 1996; Wang et al., 1998) and electrical transport (Andres et al., 1996 ; Bezryadin et al., 1997) properties of the nanostructures are affected by the shape and the size, as well as the interfeature spacing.

In general, there are two approaches to realize ordered nanostructures with desired size, shape and arrangement. One is the "bottom up" approach on pre-patterned substrates (Zhong et al, 2007; Zhong et al., 2008). The other is the "top-down" approach (Ito & Okazaki, 2000). Both of these two approaches are always based on lithographic technology. In the first approach, lithographic techniques were employed to fabricate various patterned substrates, on which ordered nanostructures can then be realized by subsequent growth of desired materials. The main reason for this approach is to suppress defects in the nanostructures. In the second approach, ordered nanostructures can be directly fabricated by lithographic techniques. Several standard lithographic techniques are frequently exploited to fabricate desired surface nanostructures, including holographic lithography, electron-beam lithography (EBL) and ion-beam lithography (IBL) (Arshak et al., 2004 ; Ebbesen et al., 1998; Ito & Okazaki, 2000). Recently, a new extreme ultraviolet (EUV) lithography was developed, which is a potential candidate for achieving critical dimensions below 100 nm (Service, 2001). In addition, there are some other lithographic techniques applied in the fabrication of nanostructures (Haynes & Van Duyne, 2001). However, fabrication of nanostructures in a regular arrangement over large areas is still a major challenge in modern nanotechnologies. There is substantial interest in developing new technologies to facilitate pattern fabrication.

2. Nanosphere lithography (NSL)

Enormous efforts have been devoted to investigate alternative nanolithography approaches. One of promising methods is nanosphere lithography (NSL) (Fuhrmann et al., 2005; Haynes & Van Duyne, 2001; Hulteen & Van Duyne, 1995; Kosiorek et al., 2004; Sinitskii et al., 2007), which is a highly accessible, low cost, parallel fabrication process capable of producing nanostructured surfaces over large areas and with high resolution. In NSL, self-assembled nanospheres can be served directly as ordered nanostructures (Park et al., 1998) or a mask for the subsequent fabrication of nanostructures, which can be realized by deposition of desired materials, or by etching on desired substrates. A rich varieties of ordered nanostructures have been achieved by NSL, such as triangular structures (Winzer et al., 1996), metallic rings (Boneberg et al., 1997), nanopillars (Weeks et al., 2004), and multilayer with modified topography (Albrecht et al., 2005), nanodots (Chen et al., 2009; Weekes et al., 2007), 3D nanostructure (Zhang et al., 2007), discs (Hanarp et al., 2003) and nanoscale crescents (Gwinner et al., 2009; Retsch et al., 2009; Vogel et al, 2011). The shape, the size and the arrangement of ordered nanostructures can be readily controlled in combination of NSL and the subsequent deposition of desired materials (Haynes & Van Duyne, 2001; Zhang et al., 2007; Vogel et al, 2011).

2.1 Main features of NSL

To obtain desired nanostructures by NSL, one monolayer of self-assembled nanospheres is always obtained first and served as a mask for the subsequent fabrication of nanostructures. The material of the nanosphere can be nanoscale polystyrene (PS), SiO_2, or polydimethylsiloxane (PDMS) (Choi et al, 2009), and etc. The shape of the resulting pattern is most often spherical. Using PDMS, the shape and the feature size of the pattern can be modulated by changing the stretching axis and ratio of the PDMS replica. The nonspherical shaped patterns, such as rectangular or elongated hexagonal shaped patterns, can then be obtained (Choi et al, 2009). An additional noteworthy feature of the PDMS is that different pattern can be produced from a single PDMS replica mold (Choi et al, 2009). In addition, binary nanospheres composed of two different-size colloidal particles can be self-assembled both in hexagonal lattices via a two-step process (Kim et al., 2009), forming binary colloid crystals (BCCs). Such BCCs may have potential applications in the fabrication of photonic crystal structures, theoretical models of phase transition, and templates of inverse structure (Kim et al., 2009). The arrangement of self-assembled nanospheres can be close-packed or non-close-packed (Vogel et al., 2011). In general, the self-assembled nanospheres are arranged in a hexagonal lattice. Using more sophisticated processes, squarely ordered array of nanospheres can also be realized, which is speculated to be metastable structures between more stable hexagonal structures (Sun et al., 2009).

In the simplest NSL, a monolayer of close-packed nanospheres in a hexagonal lattice is first obtained on the substrate, which can be served as a mask for the subsequent deposition of desired materials. For generally vertical deposition, the three-fold interstices allow deposited material to reach the substrate, giving rise to an array of triangularly shaped nanoparticles with P6mm symmetry (Haynes & Van Duyne, 2001). The perpendicular bisector of the triangular nanoparticles, a, and the interparticle spacing, d_{ip}, are proportional to the nanosphere diameter, D, which can be simply calculated by,

$$a = \frac{3}{2}(\sqrt{3} - 1 - \frac{1}{\sqrt{3}})D, \quad d_{ip} = \frac{1}{\sqrt{3}}D$$

If double layer of nanospheres are employed as the mask, both a and d_{ip} will be changed (Haynes & Van Duyne, 2001). In addition, circular shaped interstice particles are frequently obtained in the case of small nanospheres mainly because some materials are not perpendicularly deposited in the interstices, and the general hot materials can diffuse in some region. More interestingly, for angle resolved deposition, some particularly-shaped nanostructures can be realized, such as nanochain structures (Haynes & Van Duyne, 2001) and nanocrescents (Vogel et al., 2011). If the monolayer of nanospheres is non-close-packed, more complex nanostructures can be obtained by changing the incidence angle of the material vapor beam and the azimuth angle of the vapor beam with respect to the normal direction of the nanospheres mask (Zhang et al., 2007).

The critical step of NSL is to form monolayer of ordered nanospheres on desired substrates. Several methods have been developed to form regularly arranged nanospheres on substrates, including transferal coating (Weekes et al., 2007), vertically dipping coating (Choi et al., 2009), spin coating (Hulteen & Van Duyne, 1995), drop coating (Hulteen et al., 1999), and thermoelectrically cooled angle coating (Micheletto et al., 1995). All of these formation methods are based on the ability of the nanospheres to freely diffuse to seek their lowest energy configuration. The diffusion processes and the interaction among nanospheres can be influenced by chemically modifying the nanosphere surface with a negatively charged functional group such as carboxylate or sulfate. Such a modification of the surface features of nanospheres can be easily realized for polystyrene (PS) spheres (Weekes et al., 2007). The self-assembled monolayer of nanosphere masks always include a variety of defects that arise as a result of nanosphere polydispersity, site randomness, point defects (vacancies), line defects (slip dislocations), and polycrystalline domains (Haynes & Van Duyne, 2001). These defects are always remained in the finally obtained nanostructures, which will degrade the properties of the ordered nanostructures. Therefore, it is important to try to get rid of those defects in the monolayer of self-assembled nanospheres.

2.2 NSL based on transferal coating

It was found that the transferal coating is much easier in operation to obtain ordered nanospheres in large areas than the other methods. The domain size of ordered PS spheres can be up to 1 cm² (Weekes et al., 2007). The key step of the transferal coating is to self-assemble highly ordered monolayer of PS spheres at the interface between water and air (or oil). In general, the suspensions (1-10 wt%) of PS spheres in de-ionized (DI) water are diluted in a 1:1 ratio in some spreading agent, such as ethanol or methanol. Drops of the diluted suspension of PS sphere are then introduced into water surface via a tilted glass from a pipet. On contact with the water, the PS spheres immediately form a momolayer and start to assemble. The inherent mechanism for the ordering of PS spheres at a liquid interface has been studied by several groups (Aubry & Singh, 2008; Boneva et al., 2009; Larsen & Grier, 1997; Nikolaides et al., 2002; Pieranski, 1980; Trau et al., 1996; Yeh et al., 1997). A reasonable model has been provided to account for the ordering of PS spheres at the interface (Nikolaides et al., 2002). It was suggested that the ordering arrangement of PS spheres resulted from the balance between an electrostatic repulsion and an additional capillary attraction among PS spheres. The former is originated from the negative charges on PS spheres (Weekes et al, 2007). The latter is due to the deformation of liquid meniscus by electrostatic stresses at interface. Both of these two forces were associated with an electric dipolar field, which resulted from an asymmetric charge distribution on particles at the interface due to mismatch in dielectric constant of adjacent fluids. Such a creation of the attractive capillary force is crucial because spheres with diameters of less than 5 μm generally do not have sufficient weight to deform the liquid meniscus by means of gravity (Kralchevsky & Nagayama, 2000).

Considering the importance of the electrical field in self-assembling PS spheres at interface, it is natural to find ways to change the electric field to control the arrangement of PS spheres. One way is to apply an external electric field during the self-assembly of PS spheres at the interface (Aubry & Singh, 2008; Boneva et al., 2009; Nikolaides et al., 2002; Trau et al., 1996). This external electrical field will change the charge distribution on the surface of PS Spheres, leading to the change of the electric dipole field around PS spheres. It may also exert an additional electric force on the negatively charged PS spheres, which will affect the deformation of the surface. In a word, the external electric field will affect both the attraction force and the repulsion force mentioned above. It has been found that the inter-particle distance can be remarkably changed by the electric field (Nikolaides et al., 2002). As a result, the ordering of self-assembled PS spheres can be improved considerably, as shown in Fig. 1. In addition, the effect of an external electric field on the arrangement of PS spheres is more pronounced for smaller spheres.

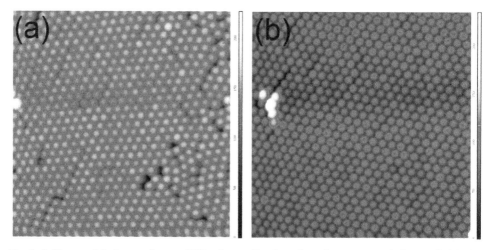

Fig. 1. Self-assembled monolayer of PS spheres (5 x 5 μm²) with an external electric field of, (a) 0 v/m, (b) 5×10^4 v/m. The diameter of PS sphere is 200 nm. The electric field is upward and perpendicular to the interface of water and air. The external electric field can efficiently reduce defects in the regular arrangement of PS spheres, remarkably improving the ordering of the self-assembled monolayer of PS spheres.

Another way to modify the balance between the attraction force and the repulsion force among PS spheres is to change the surface chemistry of PS spheres or the electrostatic environment of the water-air interface (Sirotkin et al., 2010). By adding some electrolyte, e.g. acid (H_2SO_4) and NaCl, into the water, effective surface charge density of PS spheres and /or effect of electric screen of PS spheres can be changed, which give rise to the change of the interaction among PS spheres. It has been found that the ordering of self-assembled PS spheres at interface can be considerably improved by adding suitable acid (H₂SO₄) in water (Sirotkin et al., 2010), as shown in Fig. 2. Given that the charges on PS spheres are related to the diameter of PS spheres，the suitable amount of acid or other electrolyte is dependent on the size of PS spheres。In addition, the temperature of the water to some degree also affect the self-assembly of the PS spheres at the interface. It was found that ordering of PS was improved on the water of ~ 4 ℃. Such an improvement may be related to the increase of water surface tension and the suppression of the Brownian motion of the PS spheres and dust clusters in the water.

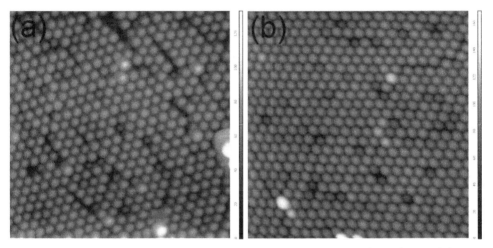

Fig. 2. Self-assembled monolayer of PS spheres (5 x 5 μm2) at the interface of air and the solution of de-ionized water and H₂SO₄ with PH value of, (a) ~7, (b) ~ 5.3. The diameter of PS sphere is 240 nm. The H₂SO₄ can provide some additional ions around PS spheres, which can effectively change the interaction among PS spheres. Under certain PH value of the solution, considerable improvement of ordering of PS spheres can be made.

The self-assembled monolayer of PS spheres can be finally transferred to varieties of smooth substrates underneath the water by draining off the water. This process can be affected by some charges on the substrates. In addition, some cracks may appear once the monolayer was disturbed by movements of the water during draining. The PS spheres nearby the cracks slightly displaced from the ideal sites of a hexagonal lattice. In this case, the long-range ordering of the subsequent structure is degraded. The monolayer of ordered PS spheres on substrates can then serve as a mask for the subsequent fabrication of ordered nanostructures by deposition of varieties of materials or by etching.

2.3 Periodic pit-pattern obtained by NSL and chemical etching

Periodic pit-pattern can be obtained in combination of NSL and selective chemical etching (Chen et al., 2009). The processes mainly involves three steps: (i) self-assembling monolayer of PS spheres on hydrogenated Si surface; (ii) forming a novel net-like Au-Oxide mask via Au catalyzed oxidation; (iii) resulting in periodic pits by selective chemical etching of Si in KOH solution.

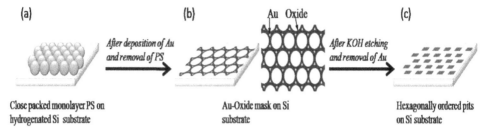

(a)   (b)   Au Oxide   (c)

After deposition of Au and removal of PS   *After KOH etching and removal of Au*

Close packed monolayer PS on hydrogenated Si substrate   Au-Oxide mask on Si substrate   Hexagonally ordered pits on Si substrate

Fig. 3. Schematic illustration for the fabrication of ordered pit-pattern using nanosphere lithography.

The experimental processes are schematically illustrated in Figure 3. It starts with self-assembling monolayer of PS spheres onto hydrogenated Si (001) substrates. The PS spheres with different diameters of 100nm, 200nm, 500nm, 600nm or 1.6 μm were used. All PS sphere suspensions were purchased from Duke Scientific Corporation. The Si (001) substrates were chemically cleaned and hydrogenated by a subsequent HF dip. The close-packed monolayer of PS spheres was self-assembled on the surface of DI water as mentioned above, which was then transferred onto Si (001) substrates immersed in DI water by draining away the DI water. Fig. 4 shows a large-area highly ordered PS spheres on a hydrogenated Si substrate.

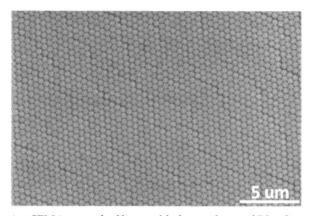

Fig. 4. Representative SEM image of self-assembled monolayer of PS spheres with a diameter of 500 nm on a hydrogenated Si (001) substrate.

The ordered PS spheres in a hexagonal lattice on Si substrate then serves as a mask for thermal evaporation of Au. After deposition of about 1 nm Au onto the PS spheres covered substrates, six Au particles around each PS spheres on Si surface were obtained. Because Au is directly deposited onto the Si surface without SiO_2, Si adjacent to Au particles then electrochemically oxidizes, or anodizes upon exposure to air (Robinson et al., 2007). As a result, an Au-oxide mask was naturally formed, as illustrated in Fig. 3(b). To avoid oxidation of the Si surface underlying PS spheres, the samples were immediately rinsed in tetrahydrofuran to remove PS spheres and then etched in 20% KOH solution at room temperature. Since Au-oxide mask protects the underlying Si from KOH etching, well ordered and uniformed 2D pit-pattern was then formed.

The periodicity and the size of the pits were determined by the diameter of PS spheres. The period of the patterned pits is in fact equal to the diameter of PS spheres. This indicates that the period of ordered pits can be readily changed by using PS with corresponding diameter, as demonstrated in Fig 5. In addition, the lateral size of inverted-pyramid-like pits was essentially decreased linearly with the diameter of PS, as shown in Fig.6. This is mainly because the area underlying the PS spheres for etching is nearly proportional to the projection of PS spheres on Si substrates. Moreover, according to the fitting line in Fig. 6, the lateral size of pits will decrease to be zero when the diameter of spheres equals to ~40nm. This result can be explained by the fact that the Au-catalyzed SiO_2 partially fill up the region underlying the PS. It means that the minimum period of patterned pits can be down to 40 nm by present method. On the other hand, any dispersion of the diameter of PS will degrade the ordering of the pits near this limit. Thus, before approaching the limit of the nanosphere lithography, more uniformed PS is required so that homogenous and ordered small pits can be obtained.

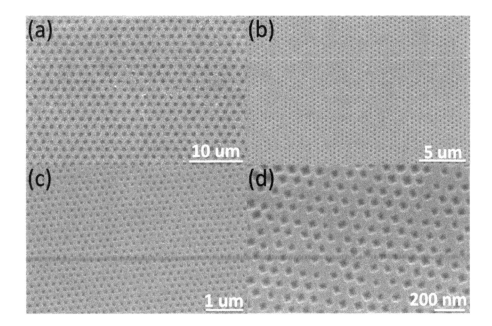

Fig. 5. Representative SEM images of pit-patterns with a periodicity of: (a) 1.6 um (after 1 minute etching), (b) 500 nm (after 3 minutes etching), (c) 200 nm (after 3 minutes etching), (d) 100 nm (after 1 minute etching）.

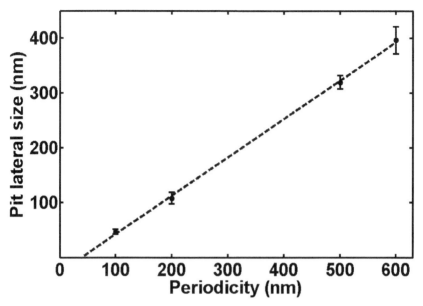

Fig. 6. Mean lateral size vs periodicities of pits. The broken curve is a fitting line.

Moreover, the shape of pits can be tuned by controlling KOH etching time. Depending on the etching time, three types of pits can be obtained. At the beginning of etching, shallow pits with rounded open-mouth (type I pits) can be obtained, as shown in Fig. 7 (a). For an intermediate etching time, inverted truncated-pyramid-like pits with {111} facets and flat (001) bottom (type II pits) are obtained, as shown in Fig. 7 (b). After sufficiently long etching time, inverted pyramid-like pits with {111} facets (type III pits) are obtained, as shown in Fig. 7 (c) and (d). Such a shape evolution with etching time is related to the anisotropic etching rate of Si by KOH solution (up to 100:1 for the etching of Si along <100> and <111> direction at room temperature). As a result, {111} facets will finally appear in the pits. The depth of type III pits approximates to be $w / 2 \times tg(54.7^{\circ})$, where w is the lateral size of pits with sidewalls of {111} facets having slope angle of 54.7°. In addition, the etching time corresponding to each type of pits depends on the periodicity of the patterned pits. As described above, a larger periodicity of patterned pits is accompanied with a larger area region for etching ' which gives rise to a larger pits. It takes time to form the sidewalls of pits completely with {111} facets from their appearance. The larger pits will have larger area sidewalls, as results in longer etching time corresponding to different types of pits. Therefore, the pits with a periodicity of 1.6 μm in Fig. 5 (a) are of type I. The pits with a periodicity of 500 nm in Fig. 5 (b) are actually of type II. The pits with a periodicity of 200 nm and 100 nm in Fig. 5 (c) and (d) are of type III. It has been found that some materials growth on Si substrates was orientation-dependent (Zhang et al., 2008). Such patterned Si substrates with the coexistence of spatially ordered (001) surface and {111} facets may provide potential templates to form ordered unique nanostructures of orientation dependence.

Well ordered GeSi nano-islands were obtained by deposition of Ge on such pit-patterned Si (001) substrates using molecular beam epitaxy (Chen et al., 2009), as shown in Fig. 8(c). Such preferential formation of GeSi nano-islands within each pit is energetically favorable under the assistance of growth kinetics (Zhong et al, 2007; Zhong et al., 2008). In comparison, GeSi nano-islands on a flat substrate under identical growth conditions are random, as shown in Fig 8(d). With decreasing the periodicity of the pit-pattern by using small PS spheres, higher density of smaller GeSi nano-islands in a hexagonal lattice are expected, which can facilitate the investigation of size-dependent quantum confinement effect of nano-islands.

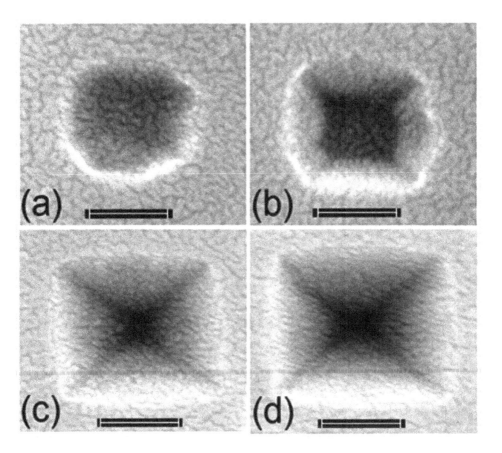

Fig. 7. SEM image of a pit (the diameter of PS spheres used is 600 nm) obtained after etching of (a) 2 minutes, (b) 3 minutes, (c) 5 minutes, (d) 7 minutes. The scale bar is 200 nm.

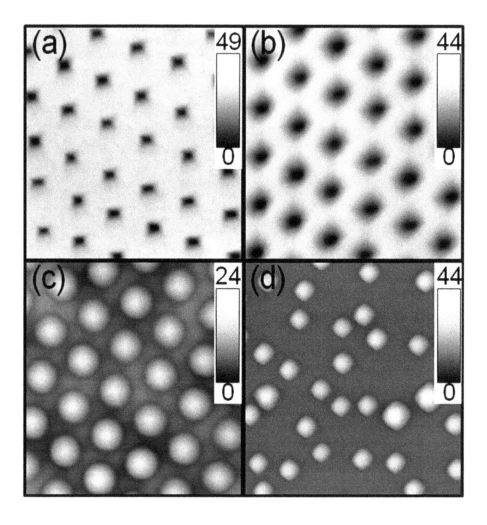

Fig. 8. AFM images (1×1 um²) of (a) pit-pattern with a periodicity of 200 nm, (b) pit-pattern after Si buffer layer growth, (c) ordered GeSi nano-islands after deposition of 10 monolayers Ge by molecular beam epitaxy on a pit-patterned Si (001) substrate, (d) Randomly distributed GeSi nano-islands grown under the same conditions on a flat Si (001) substrate. The unit of height bar is nm.

3. Conclusions

A scalable approach to fabricate periodic nanopatterning in a large-scale area with controllable periodicity using nanospheres, so called NSL, has been developed. The NSL generally started with self-assembling monolayer of PS spheres on the substrates, which can be obtained by various methods. One potential routine to obtain ordered monolayer of PS spheres is via self-assembling PS spheres at the interface between water and air. Such a regular arrangement of monolayer PS spheres in a hexagonal lattice resulted from the balance between an electrostatic repulsion among adjacent spheres and a capillary attraction due to the deformation of liquid meniscus by electrostatic stresses. An external electrical field perpendicular to the water surface, which affected the interaction between PS spheres, could efficiently improve the ordering of PS spheres, particularly of small PS spheres. The interplay among PS spheres can also be affected by changing the surface chemistry of PS spheres or the electrostatic environment of the water-air interface, which can be readily realized by mixing some electrolytes, such as solution of H_2SO_4 or NaCl. In addition, it was found that the ordering of PS spheres was improved on the water of ~ 4 °C mainly due to the increase of water surface tension and the suppression of the Brownian motion of the PS spheres and dust clusters in the water. This ordered monolayer PS spheres could be transferred onto the substrate placed previously inside water by draining off the water. This method facilitates large-area highly ordered monolayer of PS spheres on substrates, which can act as a mask or a template for subsequent lithography to obtain ordered nano-wires or nano-pits, or for subsequent growth of desired nanostructures. Two-dimensionally ordered nanopattern with a periodicity equal to the diameter of PS spheres in the range of several micrometers to less than 100 nm could be readily obtained. The geometrical profiles of the nanopattern could be modulated by controlling the etching conditions. NSL has been exploited in fabricating ordered nano-wires and nano-dots. This technique was characterized by its low-cost, high throughput, and easy manipulation for producing large-scale periodic patterns. More interestingly, NSL can be applied to obtain nanostructures of various materials on many kinds of substrates, which will facilitate the production of varieties of ordered nanostructures.

4. Acknowledgements

This work was supported by the special funds for Major State Basic Research Project No. 2011CB925600 of China.

5. References

Aizpurua, J.; Hanarp, P.; Sutherland, D. S.;K€all, M.; Bryant, G. W.; Garcia, D.; & Abajo, F. J. (2003) Optical Properties of Gold Nanorings. *Phys. Rev. Lett.* 90, 057401 (1-4)

Albrecht, M.; Hu, G. H.; Guhr, I. L.; Ulbrich, T. C.; Boneberg, J.; Leiderer, P.; & Schatz, G. (2005) Magnetic multilayers on nanospheres. *Nat. Mater.* 4, 203-206

Andres, R. P.; Bielefeld, J. D.; Henderson, J. I.; Janes, D. B.; Kolagunta, V. R.; Kubiak, C. P.; Mahoney, W. J.; & Osifchin, R. G. (1996) Self-Assembly of a Two-Dimensional Superlattice of Molecularly Linked Metal Clusters. *Science* 273, 1690-1693

Arshak, K.; Mihov, M.; Arshak, A.; McDonagh, D.; & Sutton, D. (2004) Novel dry-developed focused ion beam lithography scheme for nanostructure applications. *Microelectron. Eng.* 73/74, 144-151

Aubry, N.; & Singh, P. (2008) Physics underlying controlled self-assmbly of micro-and nanoparticles at a two-fluid interface using an electric field. *Phys. Rev. E* 77, 56302 (1-11)

Bezryadin, A.; Dekker, C.; & Schmid, G. (1997) Electrostatic trapping of single conducting nanoparticles between nanoelectrodes. *Appl. Phys. Lett.* 71, 1273-1275

Boneberg, J.; Burmeister, F.; Schafle, C.; Leiderer, P.; Reim, D.; Fery, A.; & Herminghaus, S. (1997). The Formation of Nano-Dot and Nano-Ring Structures in Colloidal Monolayer Lithography. *Langmuir* 13, 7080-7084

Boneva, M. P.; Danov, K. D.; Christov, N. C.; & Kralchevsky, P. A. (2009). Attraction between Particles at a Liquid Interface Due to the Interplay of Gravity- and Electric-Field-Induced Interfacial Deformations. *Langmuir* 25, 9129-9139

Bratlie, K. M.; Lee, H.; Komvopoulos, K.; Yang, P.; & Somorjai, G. A. (2007). Platinum Nanoparticle Shape Effects on Benzene Hydrogenation Selectivity. *Nano Lett.* 7, 3097-3101

Brolo, A. G.; Gordon, R.; Leathem, B.; & Kavanagh, K. L. (2004). Surface Plasmon Sensor Based on the Enhanced Light Transmission through Arrays of Nanoholes in Gold Films. *Langmuir* 20, 4813-4815

Chen, P.; Fan, Y.; & Zhong, Z. (2009) The fabrication and application of patterned Si(001) substrates with ordered pits via nanosphere lithography. *Nanotechnology* 20, 095303 (1-5)

Chen, X.; Wei, X.; & Jiang, K. (2008) Large-scale fabrication of ordered metallic hybrid nanostructures. *Opt. Express* 16, 11888-11893

Choi, H. K.; Im, S. H.; & Park, O (2009) Shape and Feature Size Control of Colloidal Crystal-Based Patterns Using Stretched Polydimethylsiloxane Replica Molds. *Langmuir* 25, 12011-12014

Chou, S. Y.; Wei, M. S.; Krauss, P. R.; & Fischer, P. B. (1994) Single-domain magnetic pillar array of 35 nm diameter and 65 Gbits/in.2 density for ultrahigh density quantum magnetic storage. *J. Appl. Phys.*, 76, 6673-6675

Chovin, A; Garrigue, P.; Manek-H€onninger, I.; & Sojic, N. (2004) Fabrication, Characterization, and Far-Field Optical Properties of an Ordered Array of Nanoapertures. *Nano Lett.* 4, 1965-1968

Ebbesen, T. W.; Lezec, H. J.; Ghaemi, H. F.; Thio, T.; & Wolff, P. A. (1998) Extraordinary optical transmission through sub-wavelength hole arrays. *Nature* 391, 667

Fuhrmann, B.; Leipner, H. S.; Hoche, H. R.; Schubert, L.; Werner, P.; Gösele, U. (2005) Ordered Arrays of Silicon Nanowires Produced by Nanosphere Lithography and Molecular Beam Epitaxy. *Nano Lett.* 5, 2524-2527

Gwinner, M. C.; Koroknay, E.; Fu, L. W.; Patoka, P.; Kandulski, W.; Giersig, M.; & Giessen, H. (2009) Periodic Large-Area Metallic Split-Ring Resonator Metamaterial Fabrication Based on Shadow Nanosphere Lithography. *Small* 5, 400-406

Hanarp, P.; Kall, M.; & Sutherland, D. S. (2003) Optical Properties of Short Range Ordered Arrays of Nanometer Gold Disks Prepared by Colloidal Lithography. *J. Phys. Chem. B* 107, 5768-5772

Haynes, C. L.; & Van Duyne, R. P. (2001) Nanosphere Lithography: A Versatile Nanofabrication Tool for Studies of Size-Dependent Nanoparticle Optics. *J. Phys. Chem. B* , 105, 5599-5611

Hulteen, J. C.; & Van Duyne, R. P. (1995) Nanosphere lithography: A materials general fabrication process for periodic particle array surfaces. *J. Vac. Sci. Technol. A* 13, 1553-1558

Hulteen, J. C.; Treichel, D. A.; Smith, M. T.; Duval, M. L.; Jensen,T. R.; & Van Duyne, R. P. (1999) Nanosphere Lithography: Size-Tunable Silver Nanoparticle and Surface Cluster Arrays. *J. Phys. Chem. B* 103, 3854-3863

Ito, T; & Okazaki, S. (2000) Pushing the limits of lithography. *Nature* 406, 1027-1031

Kim, J. J.;Li, Y.; Lee, E. J.; § Cho, S. O. (2009) Fabrication of Size-Controllable Hexagonal Non-Close-Packed Colloidal Crystals and Binary Colloidal Crystals by Pyrolysis Combined with Plasma–Electron Coirradiation of Polystyrene Colloidal Monolayer. *Langmuir* 27, 2334-2339

Kosiorek, A.; Kandulski, W.; Chudzinski, P.; Kempa, K.; & Giersig, M. (2004) Shadow Nanosphere Lithography: Simulation and Experiment. *Nano Lett* 4 1359-1363

Kralchevsky, P. A.; & Nagayama, K. (2000) Capillary interactions between particles bound to interfaces, liquid films and biomembranes. *AdV. Colloid Interface Sci.* 85, 145-192

Larsen, A. E.; & Grier, D. G. (1997) Like-charge attractions in metastable colloidal crystallites *Nature* 385, 230-233

Larsson, E. M.; Alegret, J.; K€all, M.; & Sutherland, D. S. (2007) Sensing characteristics of NIR localized surface plasmon resonances in gold nanorings for application as ultrasensitive biosensors. *Nano Lett.* 7, 1256-1263

Ma, X. (2008) Memory properties of a Ge nanoring MOS device fabricated by pulsed laser deposition. *Nanotechnology* 19, 275706 (1-4)

Micheletto, R.; Fukuda, H.; & Ohtsu, M. (1995) A simple method for the production of a two-dimensional, ordered array of small latex particles. *Langmuir,* 11, 3333-3336

Narayanan, R.; & El-Sayed, M. A. (2004) Shape-Dependent Catalytic Activity of Platinum Nanoparticles in Colloidal Solution. *Nano Lett.* 4, 1343-1348

Nikolaides, M. G.; Bausch, A. R.; Hsu, M. F.; Dinsmore, A. D.; Brenner, M. P.; & Weitz, D. A. (2002) Electric-field-induced capillary attraction between like-charged particles at liquid interfaces. *Nature* 420, 299-301

Park, S. H.; Qin, D.; & Xia, Y. (1998) Crystallization of Mesoscale Particles over Large Areas. *Adv. Mater.* 10, 1028-1032

Pieranski, P. (1980) Two-Dimensional Interfacial Colloidal Crystals. *Phys. ReV. Lett.* 45, 569-572

Retsch, M.; Tamm, M.; Bocchio, N.; Horn, N.; Forch, R.; Jonas, U.; Kreiter, M. (2009) Parallel Preparation of Densely Packed Arrays of 150-nm Gold-Nanocrescent Resonators in Three Dimensions. *Small* 5, 2105-2110

Robinson, J. T.; Evans, P. G.; Liddle, J. A.; & Dubon, O. D. (2007) Chemical Nanomachining of Silicon by Gold-Catalyzed Oxidation. *Nano Lett.* 7, 2009-2013

Schmidt, O. G.; & Eberl, K. (2001) Self-assembled Ge/Si dots for faster field-effect transistors. *IEEE Trans. Elec. Devices* 48, 1175-1179

Service, R. F. (2001) Optical Lithography Goes to Extremes--And Beyond. *Science* 293, 785-786

Shi, J.; Gider, S.; Babcock, K.; & Awschalom, D. D (1996) Magnetic Clusters in Molecular Beams, Metals, and Semiconductors. *Science* 271, 937-941

Sinitskii, A.; Neumeier, S.; Nelles, J.; Fischler, M.; & Simon, U. (2007) Ordered arrays of silicon pillars with controlled height and aspect ratio. *Nanotechnology* 18, 305307 (1-6)

Sirotkin, E.; Apweiler, J. D.; & Ogrin, F. Y. (2010) Macroscopic Ordering of Polystyrene Carboxylate-Modified Nanospheres Self-Assembled at the Water–Air Interface *Langmuir* 26, 10677-10683

Sun, C.; Min, W.; Linn, N. C.; Jiang, P.; Jiang, B. (2009) Large-scale assembly of periodic nanostructures with metastable square lattices. *J. Vac. Sci. Technol. B* 27, 1043-1047

Trau, M.; Saville, D. A.; & Aksay, I. A. (1996) Field-Induced Layering of Colloidal Crystals. *Science* 272, 706-709

Vogel, N; Fischer, J.; Mohammadi, R.; Retsch, M; Butt, H.; Landfester, K.; Weiss, C. K.; & Kreiter, M. (2011) Plasmon Hybridization in Stacked Double Crescents Arrays Fabricated by Colloidal Lithography. *Nano. Lett.* 11, 446-454

Volkmuth, W. D.; & Austin, R. H. (1992) DNA electrophoresis in microlithographic arrays *Nature* 358, 600-602

Volokitin, Y.; Sinzig, J.; de Jongh, L. J.; Schmid, G.; Vargaftik, M. N.; & Moiseev, I. I. (1996) Quantum-size effects in the thermodynamic properties of metallic nanoparticles. *Nature* 384, 621-623

Wang, Z. L.; Petroski, J. M.; Green, T. C.; & El-Sayed, M. A. (1998) Shape Transformation and Surface Melting of Cubic and Tetrahedral Platinum Nanocrystals. *J. Phys. Chem. B* 102, 6145-6151

Weekes, S. M.; Ogrin, F. Y.; & Murray, W. A. (2004) Fabrication of Large-Area Ferromagnetic Arrays Using Etched Nanosphere Lithography. *Langmuir* 20, 11208-11212

Weekes, S. M.; Ogrin, F. Y.; Murray, W. A.; & Keatley, P.S. (2007) Macroscopic Arrays of Magnetic Nanostructures from Self-Assembled Nanosphere Templates. *Langmuir* 23, 1057-1060

Winzer, M.; Kleiber, M.; Dix, N.; & Wiesendanger, R. (1996) Rapid communication Fabrication of nano-dot-and nano-ring-arrays by nanosphere lithography. *Appl. Phys. A* 63, 617-619

Yablonovitch, E. (1987) Inhibited Spontaneous Emission in Solid-State Physics and Electronics. *Phys. Rev. Lett.* 58, 2059-2062

Yae, S.; Tannka, H.; Kobayashi, T.; Fukumuro, N.; & Matsuda, H. (2005) Porous silicon formation by HF chemical etching for antireflection of solar cells. *Phys. Status Solidi c* 2, 3476-3480

Yeh, S.; Seul, M.; & Shraiman, B. I. (1997) Assembly of ordered colloidal aggregrates by electric-field-induced fluid flow. *Nature* 386, 57-59

Zhang, Z.; Wong, L. M.; Ong, H. G.; Wang, X. J.; Wang, J. L.; Wang, S. J.; Chen, H.; & Wu, T. (2008) Self-Assembled Shape- and Orientation-Controlled Synthesis of Nanoscale Cu3Si Triangles, Squares, and Wires. *Nano Lett.* 8, 3205-3210

Zhang,G.; Wang, D.; & Mohwald, H. (2007) Fabrication of Multiplex Quasi-Three-Dimensional Grids of One-Dimensional Nanostructures via Stepwise Colloidal Lithography. *Nano. Lett.* 7, 3410-3413

Zhong, Z.; Chen, P.; Jiang, Z.; & Bauer, G. (2008) Temperature dependence of ordered GeSi island growth on patterned Si (001) substrates. *Appl. Phys. Lett.* 93, 043106 (1-3)

Zhong, Z.; Schwinger, W.; Schäffler, F.; Bauer, G.; Vastola, G.; Montalenti, F.; & Miglio, L. (2007). Delayed Plastic Relaxation on Patterned Si Substrates: Coherent SiGe Pyramids with Dominant {111} Facets. *Phys. Rev. Lett.* 98, 176102 (1-4)

Zhu, F. Q.; Fan, D.; Zhu, X.; Zhu, J. G.; Cammarata, R. C.; Chien, C. L. (2004) Ultrahigh-Density Arrays of Ferromagnetic Nanorings on Macroscopic Areas. *Adv. Mater.* 16, 2155-2159

11

Lithography-Free Nanostructure Fabrication Techniques Utilizing Thin-Film Edges

Hideo Kaiju[1,2], Kenji Kondo[1] and Akira Ishibashi[1]
[1]Research Institute for Electronic Science, Hokkaido University
[2]PRESTO, Japan Science and Technology Agency
Japan

1. Introduction

Fabricating nanoscale patterns with sub-10 nm feature size has been an important research target for potential applications in next-generation memories, microprocessors, logic circuits and other novel functional devices. Typically, according to the International Technology Roadmap for Semiconductor (ITRS) from 2009, an 8.9 nm node device is targeted for the year 2024. To achieve this milestone, liquid immersion lithography and extreme ultraviolet (EUV) lithography can be expected to be among the most commonly used techniques for the fabrication of nanopatterns. With liquid immersion lithography using a wavelength of 193 nm and a high numerical aperture (NA), it has been demonstrated that 32 nm features can be patterned (Finders et al., 2008; Sewell et al., 2009). EUV lithography using a short wavelength of 13.5 nm and 0.3-NA exposure tool has also enabled the printing of 22 nm half-pitch lines (Naulleau et al., 2009).

On the other hand, attractive patterning techniques, such as a superlattice nanowire pattern transfer (SNAP) method (Melosh et al., 2003; Green et al., 2007), a mold-to-mold cross imprint (MTMCI) process (Kwon et al., 2005) and a surface sol-gel process combined with photolithography (Fujikawa et al., 2006), are currently proposed and pursued actively. The SNAP method, which is based on translating thin film growth thickness control into planar wire arrays, has enabled the production of molecular memories consisting of 16 nm wide titanium/silicon nanowires. The MTMCI process using silicon nanowires formed by spacer lithography, in which nanoscale line features are defined by the residual part of a conformal film on the edges of a support structure with the linewidth controlled by the film thickness, has been used to produce a large array of 30 nm wide silicon nanopillars. With the surface sol-gel process combined with photolithography, where the linewidth is determined by the thickness of coating silica layer on the resist pattern, the size reduction and the large area of sub-20 nm silica walls have been achieved.

Recently, we have proposed a double nano-baumkuchen (DNB) structure, in which two thin slices of alternating metal/insulator nano-baumkuchen are attached so that the metal/insulator stripes cross each other, as part of a lithography-free nanostructure fabrication technology (Ishibashi, 2003 & 2004; Kaiju et al., 2008; Kondo et al., 2008). The schematic illustration of the fabrication procedure is shown in Fig. 1. First, the metal/insulator spiral heterostructure is fabricated using a vacuum evaporator including a film-rolled-up system. Then, two thin slices of the metal/insulator nano-baumkuchen are cut out from the metal/insulator spiral heterostructure. Finally, the two thin slices are attached together face to face so that each stripe crosses in a highly clean environment (Ishibashi et al., 2005; Kaiju et al., 2005; Rahaman et al., 2008). Utilizing this DNB structure, we can expect to realize high density memory devices, the crossing point of which can be scaled down to ultimate feature sizes of a few nanometers thanks to their atomic-scale resolution of the film thickness determined by the rate of metal deposition, ranging from 0.01 to 1 nm/s. This DNB structure also gives a huge potential impact and importance of uniting bottom-up structures with top-down systems (Ishibashi, 2003). One element of the DNB structure is called a quantum cross (QC) device, which consists of two metal thin films (nanoribbons) having the edge-to-edge configuration as shown in Fig. 1 (Ishibashi, 2004; Kondo et al., 2006; Kaiju et al., 2008). In this QC device, the area of the crossed section is determined by the film thickness, in other words 1-20 nm thick films can produce 1×1-20×20 nm^2 nanoscale junctions. Since the vacuum evaporation has good spatial resolution of one atomic layer thickness, the junction size of QC devices could ultimately be as small as a few ångströms square ($10^{-2}nm^2$). This method offers a way to overcome the feature size limit of conventional optical lithography. When molecular-based self-assembled monolayers (SAMs), such as rotaxanes (Chen et al., 2003), catenanes (Balzani et al., 2000) and pseudorotaxanes (Pease et al., 2001), are sandwiched between the two thin metal films, QC devices can serve as novel non-volatile memory devices and switching devices. Moreover, when magnetic materials, such as Fe, Co and Ni, are used for the two thin metal films, QC devices can work as nanoscale spin injectors and tunneling magnetoresistance (TMR) devices. Among these devices, the resistance of the electrodes (thin metal films) can be reduced down to ~$k\Omega$ since the width of films can be easily controlled to the one as long as ~mm. This makes it possible to realize a highly sensitive detection for a junction resistance and to apply QC devices to high-frequency devices.

In this chapter, we present structural and electrical properties of Ni/polyethylene naphthalate (PEN) films used as electrodes in QC devices (Kaiju et al., 2009 & 2010) and current-voltage (I-V) characteristics for three types of QC devices. The three types of QC devices are as follows: (i) Ni/Ni QC devices (17 nm linewidth, 17×17 nm^2 junction area), in which two Ni thin films are directly contacted with their edges crossed (Kondo et al., 2009; Kaiju et al., 2010), (ii) Ni/NiO/Ni QC devices (24 nm linewidth, 24×24 nm^2 junction area), in which NiO thin insulators are sandwiched between two Ni thin-film edges (Kaiju et al., 2010) and (iii) Ni/poly-3-hexylthiophene (P3HT): 6, 6-phenyl C61-butyric acid methyl ester (PCBM)/Ni QC devices (16 nm linewidth, 16×16 nm^2 junction area), in which P3HT:PCBM organic molecules are sandwiched between two Ni thin-film edges (Kaiju et al., 2010; Kondo et al., 2010). In our study, we have successfully fabricated various types of QC devices with nano-linewidth and nano-junctions, which have been obtained without the use of electron-beam or optical lithography. Our method will open up new opportunities for the creation of nanoscale patterns and can also be expected as novel technique beyond conventional lithography. Furthermore, we present the calculation results of the electronic transport in Ni/organic-molecule/Ni QC devices and discussed their possibility for switching devices. According to our calculation, a high switching ratio in excess of 100000:1 can be obtained under weak coupling condition. These results indicate that QC devices fabricated using thin-film edges can be expected to have potential application in next-generation switching devices with ultrahigh on-off ratios.

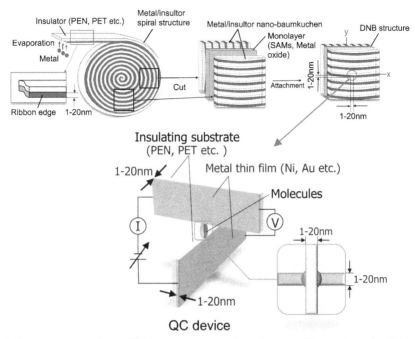

Fig. 1. Fabrication procedure of DNB structures and a schematic illustration of QC devices.

2. Fabrication and evaluation method of QC devices

2.1 Fabrication method of QC devices

The fabrication method of QC devices is shown in Fig. 2. First, Ni thin films were thermally evaporated on PEN organic substrates (2 mm width, 10 mm length and 25 μm thickness) in a high vacuum chamber at a base pressure of ~10^{-8} torr. PEN organic substrates of TEONEX Q65 were supplied by Teijin DuPont Japan and cut down from 5 to 2 mm width using a slitter in a clean environment. A boron nitride crucible N-1, made by DENKA, and a tungsten filament, made by CRAFT, were used for the thermal evaporation of Ni thin films. A heat-block stainless plate with a hole was inserted between the Ni vapor source and the PEN substrate. The length of the crucible and the aperture size in the stainless plate were designed using a geometrical simulation to evaporate uniform Ni films in-plane to PEN substrates. The temperature near PEN substrates was less than 62 °C, which was lower than the glass transition temperature T_g of 120 °C for PEN substrates. The pressure during the evaporation was 10^{-5} torr and the growth rate was 1.0-1.5 nm/min at an evaporation power of 280-350 W.

Then, fabricated Ni/PEN films were sandwiched between two polymethyl methacrylate (PMMA) resins using epoxy. The volume of the PMMA resin was 6 × 3 × 3 mm³. The edge of PMMA/Ni/PEN/PMMA structure was polished by chemical mechanical polishing (CMP) methods using alumina (Al_2O_3) slurries with particle diameters of 0.1, 0.3 and 1.0 μm. The polishing pressure was 6.5 psi and the platen rotation speed was 75 rpm. Finally, two sets of polished PMMA/Ni/PEN/PMMA structures were prepared and attached together with their edges crossed in a highly clean environment of ISO class minus 1. The attachment pressure was 0.54 MPa and no glue was used. This is a basic process in fabricating QC devices. When

Ni/Ni QC devices are fabricated, two Ni thin films are directly contacted with their edges crossed, as shown in Fig. 2(a). For the fabrication of Ni/NiO/Ni QC devices, after two sets of polished PMMA/Ni/PEN/PMMA structures were prepared, the Ni edge in one of the two samples was oxidized by O_2 plasma at a power of 5 W (= 5 mA and 1 kV) and then the oxidized and the unoxidized Ni edges were attached together with their edges crossed, as shown in Fig. 2(b). Also, for the fabrication of Ni/P3HT:PCBM/Ni QC devices, after two sets of polished PMMA/Ni/PEN/PMMA structures were prepared, P3HT:PCBM organic molecule blend was sandwiched between two sets of PMMA/Ni/PEN/PMMA structures whose edges were crossed, as shown in Fig. 2(c). P3HT and PCBM organic molecules were separately dissolved in monochlorobenzene, then blended together with a weight ratio of 1:1 to form a 20 mg/ml solution. This P3HT:PCBM solution was dripped on one of the polished PMMA/Ni/PEN/PMMA structure, then sandwiched by the other of the polished PMMA/Ni/PEN/PMMA structure. These three types of QC devices are shown in Fig. 2.

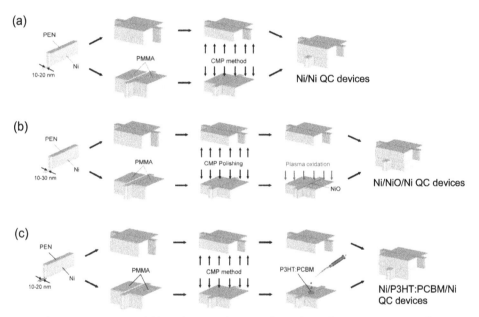

Fig. 2. Fabrication process of (a) Ni/Ni QC devices, (b) Ni/NiO/Ni QC devices and (c) Ni/P3HT:PCBM/Ni QC devices.

2.2 Evaluation methods of Ni/PEN films and QC devices

The Ni thickness was measured by a mechanical method using the stylus surface profiler DEKTAK and an optical method using the diode pumped solid state (DPSS) green laser at a wavelength of 532 nm and the photodiode detector. The surface morphologies of Ni/PEN samples were analyzed by atomic force microscope (AFM) Nanoscope IIIa. The microstructures as well as the Ni/PEN interfacial structures were examined using a JEOL JEM-3000F transmission electron microscope (TEM) operating at 300 kV. The cross-sectional TEM samples were prepared by a combination of mechanical polishing and Ar ion thinning. To reduce the beam-heating effects during ion thinning, the sample stage was cooled to -160 oC by liquid nitrogen conduction cooling. The *I-V* characteristics of QC devices were measured by a four-probe method at room temperature.

3. Ni/PEN films used as electrodes in QC devices

3.1 Cross-sectional TEM images of Ni/PEN films

Figs. 3(a) and (b) show the cross-sectional TEM images for Ni (20 nm)/PEN films. It can be seen that there is no diffusion of Ni into the PEN layer, resulting in clear and smooth formation of the Ni/PEN interface. Here, it should be noted that some researchers have reported that metal atoms diffuse into organic layers in the process of the metal evaporation onto organic layers (Tarlov, 1992; Hirose et al., 1996; Ito et al., 1999; Dürr et al., 2002). For example, the metastable atom electron spectroscopy (MAES) spectra of Au on the p-sexiphenyl (6P)/Au system shows that the features of 6P remain even though Au was deposited to about 20 nm thickness (Ito et al., 1999). This indicates that Au atoms or clusters penetrate into the 6P films. The soft x-ray photoemission spectroscopy (SXPS) investigation of the interface between evaporated indium and perylenetetracarboxylic dianhydride (PTCDA) also demonstrates that the interfacial region is very wide, ranging from 7 to 60 nm, and this means that the metal atoms of indium diffuse into PTCDA organic layers (Hirose et al., 1996). Moreover, according to studies on the interaction between evaporated Ag and octadecanethiol (ODT) on Au films using XPS, Ag deposited at 300 K migrates through the ODT layer and resides at the ODT/Au interface (Tarlov, 1992). As compared with above results, such a metal diffusion into organic layers does not occur in Ni/PEN interface. This indicates that Ni thin films on PEN organic substrates are suitable for metal/organic films used in QC devices. It can also be confirmed that the surface of Ni films is smooth, and this smoothness is in good agreement with the results of the AFM observation, where the surface roughness R_a is 1.1 nm. Fig. 3(c) shows the electron diffraction (ED) pattern for the same specimen. Ni thin films on PEN films have been shown to be face-centered-cubic structures, which are equal to those in bulk Ni structures. The ED pattern also shows that Ni thin films have polycrystalline structures, which can be recognized from the cross-sectional TEM image of Fig. 3(a). Thus, Ni /PEN films are suitable for QC devices from the viewpoint of the Ni/PEN interfacial and internal structures.

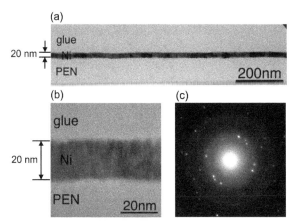

Fig. 3. (a) Cross-sectional TEM image, (b) high-resolution cross-sectional TEM image and (c) ED pattern for Ni (20 nm)/PEN films.

3.2 AFM surface morphology of Ni/PEN films

Fig. 4 shows the three-dimensional (3-D) surface images obtained from AFM observation for (a) PEN, (b) Ni (16 nm) /PEN and (c) Au (14 nm)/PEN. From the 3-D images, which are 500×500 nm^2 in area, mound-like surfaces are observed in Ni (16 nm)/PEN and Au (14 nm)/PEN, and are classified by the surface roughness R_a. Here, the surface roughness R_a is defined by

$$R_a = \frac{1}{L_x L_y} \int_0^{L_x} \int_0^{L_y} |h(x,y)|\, dxdy, \tag{1}$$

where $h(x,y)$ is the height profile as a function of x and y and $L_{x(y)}$ is the lateral scanning size in the x (y) direction. R_a of PEN is 1.3 nm, which is smaller than that of widely-used organic films, such as polyethylene terephthalate (PET) and polyimide. R_a of Ni (16 nm)/PEN is also as small as 1.22 nm. In contrast, R_a of Au (14 nm)/PEN is as large as 2.53 nm. Fig. 5 shows the surface roughness as a function of the metal film thickness for Ni/PEN and Au/PEN. R_a increases up to 3.8 nm for a film thickness of 21 nm for Au/PEN. In comparison, R_a decreases slightly down to 1.1 nm with increasing the film thickness for Ni/PEN. Here, we consider the growth mode of Ni/PEN, and discuss their feasibility in QC devices from the viewpoint of the surface roughness. Fig. 6 shows the scaling properties of R_a for PEN and Ni/PEN. The inset represents the scaling properties of the root mean square (RMS) surface roughness R_q. R_q obeys a scaling law, $R_q = w(L) \propto L^\alpha$, where $w(L)$ is the interface width corresponding to the standard deviation of the surface height, L is the system size and α is the growth scaling exponent. The growth scaling exponent for roughening, $w(L) \propto L^\alpha$, has been widely used to characterize the growth of a solid from a vapor, such as the epitaxial growth of Fe/Si (111) (Chevrier et al., 1991), growth of evaporated Ag/quartz (Palasantzas et al., 1994) and molecular beam epitaxial growth of CuCl/CaF$_2$ (111) (Tong et al., 1994), as described by the Kardar-Parisi-Zhang (KPZ) equation (Kardar et al., 1986). As for PEN and Ni/PEN, α shows the almost constant value of 0.67-0.68, as seen from the similar roughness slope in any sample. This indicates that the surface morphology of Ni/PEN exhibits almost the same behaviour as that of PEN and these results are consistent with the 3-D AFM observation in Fig. 4. We have also found that the surface is described as self-affine due to $\alpha \neq \beta$, where β is the dynamical exponent in a scaling law, $R_q = w(L) \propto t_h^\beta$. Here, t_h is a growth thickness. As one can see from Fig. 5, β is the negative value since the surface roughness slightly decreases with increasing the thickness for Ni/PEN. This results in $\alpha \neq \beta$, which shows the self-affine growth and it can also be seen in sputtered copper films (Ohkawa et al., 2002) and evaporated silver films on silicon substrates (Thompson et al., 2004). The growth process itself of Ni thin films on PEN organic substrates is of great interest and is rich in physics, so detailed work including the dynamic physical mechanism, such as the random deposition and ballistic deposition, will be reported elsewhere. Here, we consider their feasibility in QC devices from the viewpoint of the surface roughness. Since the junction area in QC devices is determined by the film thickness, we need to clarify the surface roughness in the same scanning scale as the thickness size. As shown in Fig. 6, R_a's of Ni (16 nm)/PEN and PEN are 0.34 nm and 0.44 nm, respectively, which correspond to 2-3 atomic layers, in the scanning scale of 16 nm. This result suggests that the number of molecules sandwiched between two metal thin films in QC devices can be strictly determined in a high resolution of 2-3 atoms by controlling the thickness of Ni thin films and it leads to a high product yield of memory devices and switching devices due to the reduction of the fluctuation in a junction resistance. These experimental results indicate that Ni/PEN films are suitable for QC devices from the viewpoint of the surface, interface and internal structures.

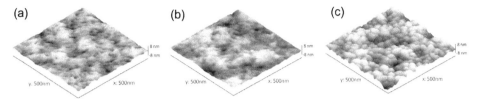

Fig. 4. 3-D surface images obtained from AFM observation for (a) PEN, (b) Ni (16 nm)/PEN and (c) Au (14 nm)/PEN.

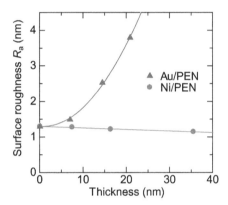

Fig. 5. Surface roughness as a function of the metal film thickness for Ni/PEN and Au/PEN.

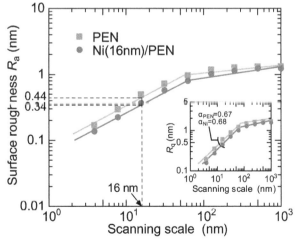

Fig. 6. Scaling properties of the surface roughness R_a for PEN and Ni/PEN. The inset represents the scaling properties of the RMS surface roughness R_q.

3.3 Electric resistivity of Ni/PEN films

Fig. 7 shows the Ni thickness dependence of the electric resistivity for Ni thin films on PEN substrates. The electric resistivity ρ_{Ni} increases with decreasing the Ni thickness d. In order to explain this experimental result quantitatively, we have calculated the electric resistivity using Mayadas-Shatzkes model (Mayadas et al., 1970). According to Mayadas-Shatzkes model, the electric resistivity ρ_{Ni} is expressed by

$$\rho_{Ni} / \rho_0 = \left[1 - \frac{3}{2}\alpha + 3\alpha^2 - 3\alpha^3 \ln(1 + \frac{1}{\alpha}) \right]^{-1}, \tag{2}$$

$$\alpha = \frac{\lambda}{D} \frac{R_g}{1 - R_g}, \tag{3}$$

where λ is the electron mean free path, D is the average grain diameter, R_g is the reflection coefficient for electrons striking the grain boundary and ρ_0 is the electric resistivity for bulk Ni. The electron mean free path λ is 11 nm for bulk Ni. The average grain diameter D is 3 nm, which has been obtained from the high-resolution TEM image and the ED pattern. The reflection coefficient R_g is 0.71-0.95, which is the extrapolation value obtained from R_g in Ni thin films with the thickness of 31-115 nm (Nacereddine et al., 2007). From Fig. 7, the experimental result shows good agreement with the calculation result quantitatively. This means that the main contribution to the electric resistivity comes from the electron scattering at grain boundaries in Ni thin films on PEN substrates. Here, we discuss the use of Ni thin films on PEN substrates for electrodes of QC devices. As can be seen from Fig. 7, the electric resistivity of Ni thin films on PEN substrates is 1-2 orders larger than that of bulk Ni. This large resistivity could produce high-resistance electrodes in QC devices. However, as stated in the introduction of this chapter, the electrode resistance can be reduced down since the film width can be controlled to the one as long as ~mm. Fig. 8(a) shows the Ni electrode resistance as a function of the linewidth l, which corresponds to the Ni thickness d, in QC devices. The schematic illustration of QC devices is shown in Fig. 8(b). In Fig. 8(a), Ni electrode resistances in the conventional cross-bar structures are also shown. The black solid line, dashed line and dotted line represent Ni electrode resistances estimated in conventional cross-bar structures with aspect ratios of 1:1, 3:1 and 5:1, respectively, where the resistivity is assumed to be the reported value in Ni thin films on glass substrates (Vries, 1987). The schematic illustration of conventional cross-bar structures is shown in Fig. 8(c). From Fig. 8(a), Ni electrode resistances in the conventional cross-bar structures are larger than 0.74, 1.2 and 3.7 MΩ in aspect ratios of 5:1, 3:1 and 1:1, respectively, for a linewidth of less than 20 nm. In contrast, QC devices show electrode resistances as small as 0.3-2 kΩ for a linewidth of 10-20 nm. This resistance reduction in Ni electrodes makes it possible to detect the resistance of sandwiched materials between two edges of Ni thin films very strictly and precisely and also this result indicates that QC devices have potential application in high-frequency devices. Thus, Ni/PEN films are suited for electrodes in QC devices from the viewpoint of the electrical properties as well as the surface, interface and internal structures.

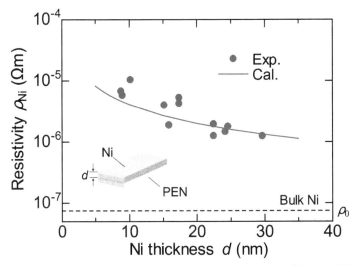

Fig. 7. Ni thickness dependence of the electric resistivity for Ni thin films on PEN substrates.

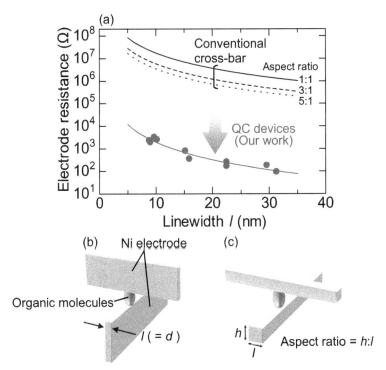

Fig. 8. (a) Ni electrode resistance as a function of the linewidth l, which corresponds to the Ni thickness d, in QC devices. Schematic illustration of (b) QC devices and (c) conventional cross-bar structures.

4. Electrical properties of QC devices

4.1 I-V characteristics of Ni/Ni QC devices

Fig. 9 shows the I-V characteristics for Ni/Ni QC devices fabricated using Ni/PEN films discussed above. The inset shows the experimental setup for the four-probe method. Since the Ni thickness is 17 nm, the cross-sectional area between two Ni thin films is 17×17 nm^2. As seen from Fig. 9, the ohmic I-V characteristics have been obtained for both positive and negative bias at room temperature. Fig. 10 shows the aging properties for the voltage with constant currents of 0.1, 0.2, 0.3, 0.5, 0.7 and 1.0 μA, respectively. The voltage is stable in any current and the standard deviation ΔV of the voltage is 22-25 mV, which corresponds to the signal-to-noise (SN) ratio of 34-52 dB, where the SN ratio is defined by 20 $\log V/\Delta V$. Here, it should be noted that the fabrication of nanojunctions using the film edges had been challenged by other researchers before (Nawate, et al., 2004). According to their attempts, Co and Ni thin films were evaporated onto glass substrates using vacuum evaporation and then they were cleaved and stuck to each other with their edges crossing. Although the current flowed across the junction, there remained a few problems: that the edge angle had to be inclined at a 15-25° and the film thickness had to be larger than 50 nm. Furthermore, the current was slightly changed as time passed although the current flowed. In contrast, such problems have not occurred in our experiments, and we have obtained stable ohmic characteristics, where there has been no change with time, for the 17 nm size junction. These experimental results indicate that our method using thin-film edges can be expected to work as a new nanostructure fabrication technology beyond conventional lithography.

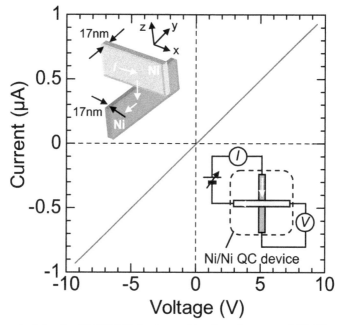

Fig. 9. I-V characteristics for Ni/Ni QC devices with a junction area of 17×17 nm^2. The inset shows the experimental setup for the four-probe method.

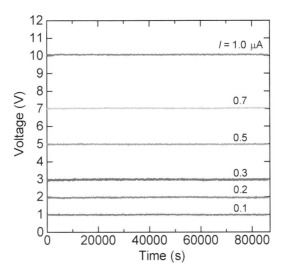

Fig. 10. Aging properties for Ni/Ni QC devices with a junction area of 17×17 nm^2.

4.2 J-V characteristics of Ni/NiO/Ni QC devices

In this section, we present current density-voltage (J-V) characteristics in Ni/NiO/Ni QC devices, which consist of NiO tunnel barriers sandwiched between two Ni thin films whose edges are crossed. First, we introduce the background of sub-micrometer scale tunnel junctions and the motivation for fabricating QC devices with tunnel barriers. Then, we show the derivation of a formula for J-V characteristics of QC devices with tunnel barriers and finally demonstrate experimental results of their J-V characteristics.

4.2.1 Background of sub-micrometer scale tunnel junctions

Sub-micrometer scale tunnel junctions have attracted much interest due to their potential application in magnetic random access memories (MRAMs), fast detectors of terahertz (THz) and IR radiation and superconducting quantum interference devices (SQUIDs). Magnetic tunnel junctions (MTJs), which consist of two ferromagnetic metals separated by thin insulators, can be expected as ultrahigh-density MRAM devices because of the giant TMR effect at room temperature (Miyazaki et al., 1995; Moodera et al., 1995; Yuasa et al., 2004; Parkin et al., 2004). CoFeB/MgO/CoFeB MTJs with a small junction area of 0.02 μm^2 exhibit a large TMR ratio of 98 %, where a clear current-induced magnetization switching (CIMS) with a low switching current density of 3.6 MA/cm^2 have been observed (Hayakawa et al., 2008). Antenna-coupled tunnel junction devices, which consist of metal/insulator/metal tunnel junctions coupled to a thin-film metal antenna, can also be expected as fast detectors of THz and IR radiation (Sanchez et al., 1978; Kale, 1985; Hobbs et al., 2005). Ni/NiO/Ni tunnel junctions with a junction area of 0.16 μm^2 coupled to thin-film metal antennas can serve as IR detectors and frequency mixers in the 10 μm band (Wilke et al., 1994; Fumeaux et al., 1996). Moreover, much effort has been devoted to the development of sub-micrometer scale SQUIDs, which are very promising devices with a high magnetic flux sensitivity (Rugar et al., 2004; Troeman et al., 2007; Huber et al., 2008; Kirtley et al., 2009). Aluminum SQUIDs with an effective area of 0.034 μm^2 display a high flux sensitivity of 1.8×10⁻⁶ $\Phi_0/Hz^{1/2}$, where $\Phi_0 = h/2e$ is the flux quantum, and operates in fields as high as 0.6 T (Finkler et al., 2010). Thus, sub-micrometer scale tunnel junctions can find a wide range application in various functional electronic devices.

As shown in the previous section, our nanostructure fabrication method using thin-film edges offers a way to overcome the feature size limit of conventional lithography. This method can be extended to fabricating nanoscale tunnel junctions. The realization of nanoscale tunnel junctions leads to the enhancement of the performance in MRAMs, fast detectors of THz and IR radiation and SQUIDs. From this motivation, it is of great importance to fabricate QC devices with tunnel barriers and very meaningful to investigate their J-V characteristics.

4.2.2 Derivation of a formula for J-V characteristics of QC devices with tunnel barriers

We derive a formula for J-V characteristics in QC devices with tunnel barriers, which consist of thin insulating barriers sandwiched between two thin metal films whose edges are crossed, shown in Fig. 11. The thicknesses of the top and bottom metal films are l_z and l_y, respectively. The barrier height and the barrier thickness are ϕ and d, respectively. The number of electrons per unit area and unit time from the top to bottom metal is given by

$$N_S^T = \int_0^\infty v_x n(v_x) |T(v_x)|^2 \, dv_x$$
$$= \frac{1}{2\pi} \int_0^\infty \frac{1}{\hbar} \frac{\partial E}{\partial k_x} n(k_x) |T(k_x)|^2 \, dk_x \,,$$

(4)

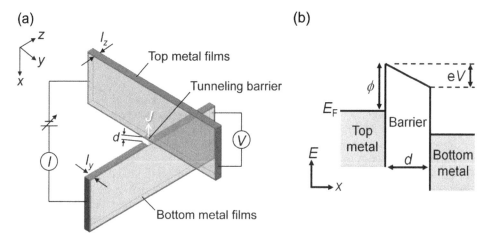

Fig. 11. (a) QC devices with tunnel barriers and (b) the potential profile of the barrier, which is used for the calculation of the transmission probability.

where v_x is the x-direction velocity, $n(v_x)$ is the number of electrons per unit volume and unit velocity, k_x is the x-direction wave number, $n(k_x)$ is the electron density, $\hbar$ is Planck's constant divided by 2π and E is the total energy of electrons, which is expressed by

$$E = E_x + E_y + E_z$$

$$= \frac{\hbar^2}{2m}\{k_x^{\,2} + k_y^{\,2} + (\frac{n\pi}{l_z})^2\}, \tag{5}$$

where E_x, E_y and E_z are the x, y and z components of the total energy of electrons, respectively, m is the electron mass, k_y is the y-direction wave number and n is the positive integer. $T(k_x)$ is the transmission probability of electrons through the barrier. $T(k_x)$ is calculated by the well-known Wentzel-Kramers-Brillouin (WKB) approximation. The potential profile is illustrated in Fig. 11(b). Here, the electron density is given by

$$n(k_x) = \frac{1}{2\pi^2}\int_0^\infty \int_0^\infty f(E)dk_y dk_z , \tag{6}$$

where $f(E)$ is the Fermi-Dirac distribution function. Substituting Eqs. (5) and (6) in Eq. (4) yields

$$N_S^T = \frac{1}{4\pi^3\hbar}\int_0^\infty |T(k_x)|^2 [\int_0^\infty \int_0^\infty f(E)dk_y dk_z]dE_x. \tag{7}$$

Since the density of states in the k_y direction for the top metal is equal to the solution of one-dimensional density of states, the relation between k_y and E_y is given by

$$dk_y = \frac{m}{\hbar\sqrt{2mE_y}}dE_y. \tag{8}$$

Substituting Eqs. (5) and (8) in Eq. (7) gives

$$N_S^T = \frac{1}{4\pi^3\hbar}\int_0^\infty |T(k_x)|^2 [\sum_{n=1}^\infty \int_0^\infty f(E)\frac{m}{\hbar\sqrt{2mE_y}}\theta(E_y - \frac{\hbar^2}{2m}(\frac{n\pi}{l_z})^2)dE_y]dE_x, \tag{9}$$

where $\theta(x)$ is the unit step function. The number of electrons per unit area and unit time from the bottom to top metal is also derived in a similar manner. When the positive bias V is applied to the bottom metal, the Fermi-Dirac distribution function is written by $f(E+eV)$. Thus, the number of electrons from the bottom to top metal is

$$N_S^B = \frac{1}{4\pi^3\hbar}\int_0^\infty |T(k_x)|^2 [\sum_{n=1}^\infty \int_0^\infty f(E+eV)\frac{m}{\hbar\sqrt{2mE_z}}\theta(E_z - \frac{\hbar^2}{2m}(\frac{n\pi}{l_y})^2)dE_z]dE_x. \tag{10}$$

Consequently, we obtain for the net current density J through the barrier

$$J = e(N_S^T - N_S^B)$$

$$= \frac{e}{4\pi^3\hbar}\int_0^\infty |T(k_x)|^2 [\sum_{n=1}^\infty \int_0^\infty f(E)\frac{m}{\hbar\sqrt{2mE_y}}\theta(E_y - \frac{\hbar^2}{2m}(\frac{n\pi}{l_z})^2)dE_y \tag{11}$$

$$- \sum_{n=1}^\infty \int_0^\infty f(E+eV)\frac{m}{\hbar\sqrt{2mE_z}}\theta(E_z - \frac{\hbar^2}{2m}(\frac{n\pi}{l_y})^2)dE_z]dE_x.$$

We can calculate J-V characteristics as a function of the metal thickness, the barrier height and the barrier thickness using Eq. (11).

4.2.3 J-V characteristics of Ni/NiO/Ni QC devices

Fig. 12 shows the J-V characteristics for Ni/NiO/Ni QC devices at room temperature. Since the Ni thickness is 24 nm, the junction area is 24×24 nm². In Fig. 12(a), solid circles are experimental data and dashed, solid and dotted lines are calculation results for barrier heights of 0.6, 0.8 and 1.0 eV, respectively, with a constant barrier thickness of 0.63 nm. In Fig. 12(b), solid circles are experimental data and dashed, solid and dotted lines are calculation results for barrier thicknesses of 0.42, 0.63 and 0.84 nm, respectively, with a barrier height of 0.8 eV. As seen from Fig. 12, experimental results are in good agreement with calculation results for a barrier height of 0.8 eV and a barrier thickness of 0.63 nm. Since the crystal structure of NiO is a NaCl-type structure with a lattice constant a = 0.42 nm, 2 monolayer of NiO is formed as shown in the inset of Fig. 12. Here, we discuss the barrier height of NiO insulating layers. The bulk NiO is a charge-transfer insulator, which has been approved by X-ray absorption (Kuiper et al., 1989) and X-ray photoemission and bremsstrahlung isochromat spectroscopy (Sawatzky et al., 1984; Elp et al., 1992). According

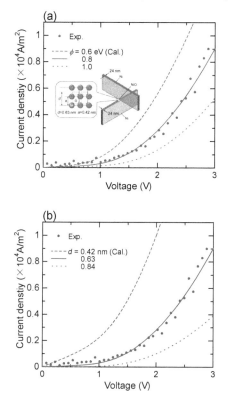

Fig. 12. (a) Barrier height dependence and (b) barrier thickness dependence of J-V characteristics for Ni/NiO/Ni QC devices with a junction area of 24×24 nm² at room temperature.

to these experiments, the energy gap is 4.0-4.3 eV, which agrees with recent calculation results performed using the full-potential linearized augmented plane wave (FLAPW) method within the LSDA(GGA)+U (LSDA, local-spin-density approximation; GGA, generalized gradient approximation; U, on-site Coulomb interation) (Cai et al., 2009). Since the Fermi energy of NiO is located in the level of 1.0 eV higher than the top of the valence band (Sawatzky et al., 1984), the barrier height in metal/NiO/metal tunnel junctions is estimated to be 3.0-3.3 eV. However, several researchers have reported that the barrier height in metal/NiO/metal tunnel junctions tends to become smaller than the estimated value of 3.0-3.3 eV (Doudin et al., 1997; Ono et al., 1997; Hobbs et al., 2005). For example, the barrier height of Ni/NiO/Ni tunnel junctions with a junction area of 0.3 μm^2 is as low as 0.18-0.22 eV with a barrier thickness of 2.5 nm (Hobbs et al., 2005). Electrodeposited Ni/NiO/Co tunnel junctions with a cross section of 0.01 μm^2 exhibit a low barrier height of 0.2-0.4 eV with a barrier thickness of 2-3 nm (Doudin et al., 1997). Thus, the barrier height of NiO thin films is reported to be smaller than that of NiO bulk materials. In our experiments, the barrier height in Ni/NiO/Ni QC devices is also estimated to be as low as 0.8 eV. This means that the derived formula is useful for the evaluation of the barrier height and the barrier thickness in QC devices with tunnel barriers. This fact also indicates that high-quality NiO thin films were formed and Ni/NiO/Ni QC devices worked well as nanoscale tunnel devices. Therefore, our method to fabricate nanoscale junctions utilizing thin-film edges can be expected to work as a new technique for the creation of nanoscale tunnel junctions.

4.3 I-V characteristics of Ni/P3HT:PCBM/Ni QC devices

In this section, we present I-V characteristics in Ni/P3HT:PCBM/Ni QC devices, which consist of P3HT:PCBM organic molecules sandwiched between two Ni thin films whose edges are crossed. First, we introduce the background of molecular electronic devices and the motivation for fabricating QC devices with organic molecules. Then, we show the theoretical calculation for I-V characteristics of QC devices with organic molecules and compare them with experimental results. Finally, we discuss their possibility for switching devices with high on-off ratios.

4.3.1 Background of molecular electronic devices

Molecular electronics have stimulated considerable interest as a technology that offers the prospect of scaling down device dimensions to a few nanometers and that also promotes a practical introduction to high-density memory applications (Chen et al., 1999; Reed et al., 2001; Chen et al., 2003; Lau et al., 2004; Wu et al., 2005; Mendes et al., 2005; Green et al., 2007). Especially, in the ITRS 2009 edition, molecular memory devices have been expected as candidates for beyond-CMOS devices since they offer the possibility of nanometer-scale components. For examples, Au/2'-amino-4-ethynylphenyl-4'-ethynylphenyl-5'-nitro-1-benzenethiolate/Au molecular devices with a junction area of 30×30-50×50 nm^2 have been fabricated using nanopore methods, and have exhibited a negative differential resistance and a large on-off peak-to-valley ratio (Chen et al., 1999). Molecular devices, which comprise a monolayer of bistable [2]rotaxanes sandwiched between two Ti/Pt metal electrodes with a linewidth of 40 nm, have also been fabricated using nanoimprint lithography, and have shown bistable I-V characteristics with high on-off ratios and reversible switching properties (Chen et al., 2003). Moreover, cross-bar molecular devices consisting of a monolayer of bistable [2]rotaxanes sandwiched between top Ti and bottom Si electrodes (16 nm wide at 33 nm pitch) have been fabricated using SNAP methods, and have acted as 160 kbit memories patterned at a density of 100 Gbit/cm² (Green et al., 2007). Thus, molecular electronic devices can be expected as next-generation high-density memories.

As demonstrated in the previous section, our method using thin-film edges provides the fabrication of nanoscale tunnel junctions. This method can be extended to fabricating nanoscale molecular devices. The realization of nanoscale molecular devices leads to the application in next-generation high-density memories. From this motivation, it is of great significance to fabricate QC devices with organic molecules and very interesting to investigate their J-V characteristics.

4.3.2 Theoretical calculation and experimental results of I-V characteristics in Ni/P3HT: PCBM/Ni QC devices

We calculate the I-V characteristics of QC devices with the molecule within the framework of Anderson model (Kondo et al., 2008 & 2009; Kondo, 2010). QC devices with the molecule consist of the molecule sandwiched between two thin metal films whose edges are crossed. The molecule is assumed to have two energy levels. In this calculation model, the Anderson Hamiltonian is given by

$$H = H_{\text{electrodes}} + H_{\text{molecules}} + H_{\text{t}}. \tag{12}$$

(a)

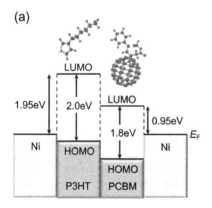

(b)

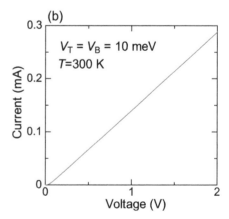

Fig. 13. (a) Energy diagram used in the calculation of I-V characteristics for QC devices and (b) the calculated I-V characteristics for QC devices under the strong coupling limit.

Here, $H_{electrodes}$ is the Hamiltonian of both of the metal electrodes, $H_{molecules}$ is the Hamiltonian of the molecule sandwiched between the metal electrodes and H_t is the transfer Hamiltonian between the sandwiched molecule and each electrode. $H_{electrodes}$, $H_{molecules}$ and H_t can be expressed by

$$H_{electrodes} = \sum_{\alpha=T,B}\sum_{k,\sigma}\varepsilon_{k\sigma}c_{\alpha,k\sigma}^{+}c_{\alpha,k\sigma},$$

$$H_{molecules} = \sum_{i,\sigma}\varepsilon_0(i)\,a_{i,\sigma}^{+}a_{i,\sigma}, \tag{13}$$

$$H_t = \sum_{\alpha=T,B}\sum_{k,\sigma}\sum_{i,\sigma}(V_{\alpha}c_{\alpha,k\sigma}^{+}a_{i,\sigma} + h.c.).$$

Here, $\varepsilon_{k\sigma}$ is the free electron energy of $\hbar^2 k^2 / 2m$, where m is the free electron mass, $\hbar$ is Planck's constant divided by 2π and k is a two-dimensional wavenumber. $c_{\alpha,k\sigma}^{+}$ and $c_{\alpha,k\sigma}$ are creation and annihilation operators for electrons of wavenumber k and spin index σ in α electrode. α indicates the top or bottom metal electrode. $\varepsilon_0(i)$ represents the ith energy level of eigenstates of the molecule. $a_{i,\sigma}^{+}$ and $a_{i,\sigma}$ are creation and annihilation operators for electrons of spin index σ in the ith energy level of the molecule. V_α is the transfer matrix between the molecule and the α electrode. In this calculation, we assume the molecule has two energy levels of $\varepsilon_0(1)=0.95$ eV and $\varepsilon_0(2)=1.95$ eV, which are estimated from Fermi levels E_F of the metal electrode, imagining that QC devices consist of the Ni electrodes and the sandwiched molecule of a P3HT:PCBM organic molecule, as shown in Fig. 13(a) (Eastman, 1970; Thompson et al., 2008). E_F is assumed to be 9.071 eV for Ni electrodes (Wang et al., 1977). Considering H_t as a perturbation, we can derive a formula for the I-V characteristics from the top to the bottom electrode using the many-body perturbation technique. As a result, the current flowing from the top to the bottom electrode can be expressed by

$$I = \frac{2e^2}{h}\int_{E_F}^{E_F+eV} d\varepsilon\sum_i\left(\frac{4\Gamma_T(\varepsilon)\Gamma_B(\varepsilon)}{(\varepsilon-\varepsilon_0(i))^2+(\Gamma_T(\varepsilon)+\Gamma_B(\varepsilon))^2}\right)[f(\varepsilon-eV-E_F)-f(\varepsilon-E_F)], \tag{14}$$

where e is the elementary charge and $f(\varepsilon)$ is the Fermi-Dirac distribution function. $\Gamma_{T(B)}(\varepsilon)$ is the coupling strength between the top (bottom) Ni electrode and the P3HT:PCBM organic molecule, which is given by

$$\Gamma_{T(B)}(\varepsilon) = \pi D_{T(B)}(\varepsilon)\,|\,V_{T(B)}|^2, \tag{15}$$

where $D_{T(B)}(\varepsilon)$ is a density of states of electrons for the top (bottom) Ni electrode and $V_{T(B)}$ is the coupling constant between the top (bottom) Ni electrode and the P3HT:PCBM organic molecule. Fig. 13(b) shows the calculated I-V characteristics for Ni/P3HT:PCBM/Ni QC devices under the strong coupling limit. $V_{T(B)}$ is assumed to be 10.0 meV, corresponding to $\Gamma_{1(B)}$ of 3927 meV. We have obtained the ohmic I-V characteristics with a resistance of 6.7 kΩ. Fig. 14 shows the experimental results of I-V characteristics for Ni/P3HT:PCBM/Ni QC devices at room temperature. The inset represents the experimental setup. The Ni thickness is 16 nm. Therefore, the junction area is 16×16 nm². We have obtained ohmic characteristics with a junction resistance of 32 Ω. Here, we compare this experimental value with the calculation result. The calculation result has demonstrated that the resistance is 6.7 kΩ, where the junction area is 1×1 nm², which is expected as a size of one P3HT:PCBM organic molecule. The number of the conductance channel is four, taking into consideration the spin degeneracy. On the other hand, in experiments, the junction area of P3HT:PCBM organic molecules is 16×16 nm², which corresponds to 1024 (=4×16×16) conductance channels.

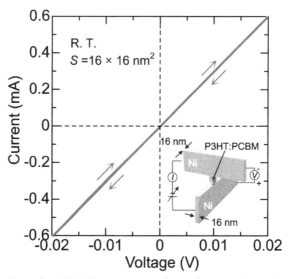

Fig. 14. Experimental results of I-V characteristics for Ni/P3HT:PCBM/Ni QC devices with a junction area of 16×16 nm² at room temperature. The inset represents the experimental setup.

Therefore, the junction resistance in a size of 16×16 nm² is calculated to be 26 Ω (=6.7kΩ/16/16), which is in good agreement with the experimental value of 32 Ω. This result indicates that electrons in nanoscale junctions can transport through the molecules in the ballistic regime without any scattering. This also demonstrates that our method to fabricate nanoscale junctions utilizing thin-film edges can be a useful tool for the creation of nanoscale molecular devices.

4.3.3 Possibility of Ni/P3HT:PCBM/Ni QC devices for switching devices

Finally, we have discussed the possibility of Ni/P3HT:PCBM/Ni QC devices for switching devices with high on-off ratios. Fig. 15 shows the calculated I-V characteristics for Ni/P3HT:PCBM/Ni QC devices under the weak coupling condition. $V_{T(B)}$ is assumed to be 0.2 meV, corresponding to $\Gamma_{T(B)}$ of 1.57 meV. From Fig. 15, the calculated result shows the sharp steps at the positions of the energy level of the P3HT:PCBM organic molecule. The off-state current I_0 is 3.8 pA at the voltage V_0 of 0.03 V, and the on-state current I_1 is 0.57 μA at the voltage V_1 of 1.03 V. As we estimate the switching on-off ratio, the I_1/I_0 ratio is found to be in excess of 100000:1. Here, it should be noted that it is essentially important that the junction area is as small as nanometer scale in order to obtain such a high on-off ratio. When the junction area is as large as micrometer scale, the number of molecules sandwiched between the electrodes is large, so the energy level can be broadened. In contrast, when the junction area is as small as nanometer scale, the number of molecule is small, so the energy level can be discrete. This discrete energy level leads to the sharp steps in the *I-V* characteristics, which can produce such a high switching ratio. Thus, Ni/P3HT:PCBM/Ni QC devices utilizing thin-film edges can be expected to have potential application in switching devices with high on-off ratios.

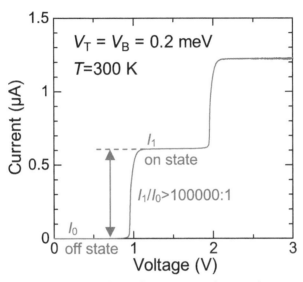

Fig. 15. Calculated *I-V* characteristics for Ni/P3HT:PCBM/Ni QC devices under the weak coupling condition.

5. Conclusions

In this chapter, we have introduced structural and electrical properties of Ni/PEN films used as electrodes in QC devices and *I-V* characteristics for three types of QC devices. The three types of QC devices are as follows: (i) Ni/Ni QC devices (17 nm linewidth, 17×17 nm² junction area), in which two Ni thin films are directly contacted with their edges crossed, (ii) Ni/NiO/Ni QC devices (24 nm linewidth, 24×24 nm² junction area), in which NiO thin insulators are sandwiched between two Ni thin-film edges and (iii) Ni/P3HT:PCBM/Ni QC devices (16 nm linewidth, 16×16 nm² junction area), in which P3HT:PCBM organic molecules are sandwiched between two Ni thin-film edges. In our study, we have successfully fabricated various types of QC devices with nano-linewidth and nano-junctions, which have been obtained without the use of electron-beam or optical lithography. Our method will open up new opportunities for the creation of nanoscale patterns and can also be expected as novel technique beyond conventional lithography. Furthermore, we have presented the calculation results of the electronic transport in Ni/organic-molecule/Ni QC devices and discussed their possibility for switching devices. According to our calculation, a high switching ratio in excess of 100000:1 can be obtained under weak coupling condition. These results indicate that QC devices fabricated using thin-film edges can be expected to have potential application in next-generation switching devices with ultrahigh on-off ratios.

6. Acknowledgment

This research has been partially supported by the Management Expenses Grants for National Universities Corporations and a Grant-in-Aid for Young Scientists from The Ministry of Education, Culture, Sports, Science and Technology (MEXT), Precursory Research for Embryonic Science and Technology program from Japan Science and Technology Agency (JST) and a Grant-in-Aid for Scientific Research from Japan Society for the Promotion of Science (JSPS). The authors would like to express their sincere appreciation to Dr. M. Hirasaka of Teijin Ltd. and Research Manager K. Kubo of Teijin DuPont Films Japan Ltd. for supplying PEN organic films, Prof. Y. Hirotsu, Assoc. Prof. M. Ishimaru and Assist. Prof. A. Hirata at Osaka University for investigating cross-sectional TEM images and ED patterns, Y. Inoue of Meiwafosis Co., Ltd. for examining the fabrication of NiO insulators and Prof. M. Yamamoto, Assist. Prof. K. Matsuda, A. Ono, N. Basheer, N. Kawaguchi, S. White, H. Sato and M. Takei at Hokkaido University for their cooperation and helpful discussions.

7. References

Balzani, V., Credi, A., Mattersteig, G., Matthews, O. A., Raymo, F. M., Stoddart, J. F., Venturi, M.; White, A. J. P. & Williams, D. J. J. (2000). *Org. Chem.* 65: 1924.

Cai, T., Han, H., Yu, Y., Gao, T., Du, J. & Hao, L. (2009). *Physica B* 404: 89.

Chen, J., Reed, M. A., Rawlett, A. M. & Tour, J. M. (1999). *Science* 286: 1550.

Chen, Y., Ohlberg, D. A. A., Li, X., Stewart, D. R., Williams, R. S., Jeppesen, J. O., Nielsen, K. A., Stoddard, J. F., Olynick, D. L. & Anderson, R. (2003). *Appl. Phys. Lett.* 82: 1610.

Chevrier, J., Lethanh, V., Buys, R. & Derrien, J. (1991). *Europhys. Lett.* 16: 737.

Doudin, B., Redmond, G., Gilbert, S. E. & Ansermet, J. –Ph. (1997). *Phys. Rev. Lett.* 79: 933.

Dürr, A. C., Schreiber, F., Kelsch, M., Carstanjen, H. D. & Dosch, H. (2002). *Adv. Mater.* 14: 961.

Eastman, D. E. (1970). *Phys. Rev. B* 2: 1.

Elp, J., Eskes, H., Kuiper, P. & Sawatzky, G. A. (1992). *Phys. Rev. B* 45: 1612.

Finders, J., Dusa, M., Vleeming, B., Megens, H., Hepp, B., Maenhoudt, M., Cheng, S. & Vandeweyer, T. (2008). *Proc. SPIE* 6924: 692408.

Finkler, A., Segev, Y., Myasoedov, Y., Rappaport, M. L., Ne'eman, L., Vasyukov, D., Zeldov, E., Huber, M. E., Martin, J. & Yacoby, A. (2010). *Nano Lett.* 10: 1046.

Fujikawa, S., Takaki, R. & Kunitake, T. (2006). *Langmuir* 22: 9057.

Fumeaux, C., Herrmann, W., Rothuizen, H., Natale, P. D. & Kneubuehl, F. K. (1996). *Appl. Phys. B* 63: 135.

Green, J. E., Choi, J. W., Boukai, A., Bunimovich, Y., Johnston-Halperin, E., Delonno, E., Luo, Y., Sheriff, B. A., Xu, K., Shin, Y. S., Tseng, H.-R., Stoddart, J. F. & Heath, J. R. (2007). *Nature* 445: 414.

Hayakawa, J., Ikeda, S., Miura, K., Yarnanouchi, M., Lee, Y. M., Sasaki, R., Ichimura, M., Ito, K., Kawahara, T., Takemura, R., Meguro, T., Matsukura, F., Takahashi, H., Matsuoka, H. & Ohno, H. (2008). *IEEE Trans. Mag.* 44: 1962.

Hirose, Y., Kahn, A., Aristov, V. & Soukiassian, P. (1996). *Appl. Phys. Lett.* 68: 217.

Hobbs, P. C. D., Laibowitz, R. B. & Libsch, F. R. (2005). *Appl. Opt.* 44: 6813.

Huber, M. E., Koshnick, N. C., Bluhm, H., Archuleta, L. J., Azua, T., Bjornsson, P. G., Gardner, B. W., Halloran, S. T., Lucero, E. A. & Moler, K. A. (2008). *Rev. Sci. Instrum.* 79: 053704.

Ishibashi, A. (2003). Jpn. Pat. 3974551.

Ishibashi, A. (2004). *Proceedings of International Symposium on Nano Science and Technology*, pp. 44-45, Tainan, Taiwan, November 20-21, 2004.

Ishibashi, A., Kaiju, H., Yamagata, Y. & Kawaguchi, N. (2005). *Electron. Lett.* 41: 735.

Ito, E., Oji, H., Furuta, M., Ishii, H., Oichi, K., Ouchi, Y. & Seki, K. (1999). *Synthetic Metals* 101: 654.

Kaiju, H., Kawaguchi, N. & Ishibashi, A. (2005). *Rev. Sci. Instrum.* 76: 085111.

Kaiju, H., Ono, A., Kawaguchi, N. & Ishibashi, A. (2008). *Jpn. J. Appl. Phys.* 47: 244.

Kaiju, H., Ono, A., Kawaguchi, N. & Ishibashi, A. (2008). *J. Appl. Phys.* 103: 07B523.

Kaiju, H., Ono, A., Kawaguchi, N., Kondo, K., Ishibashi, A., Won, J. H., Hirata, A., Ishimaru, M. & Hirotsu, Y. (2009). *Appl. Surf. Sci.* 255: 3706.

Kaiju, H., Kondo, K., Ono, A., Kawaguchi, N., Won, J. H., Hirata, A., Ishimaru, M., Hirotsu, Y. & Ishibashi, A. (2010). *Nanotechnology* 21: 015301.

Kaiju, H., Kondo, K., Basheer, N., Kawaguchi, N., White, S., Hirata, A., Ishimaru, M., Hirotsu, Y. & Ishibashi, A. (2010). *Mater. Res. Soc. Symp. Proc.* 1252: J0208.

Kaiju, H., Basheer, N., Kondo, K. & Ishibashi, A. (2010). *IEEE Trans. Magn.* 46: 1356.

Kaiju, H., Kondo, K. & Ishibashi, A. (2010). *Jpn. J. Appl. Phys.* 49: 105203.

Kale, B. M. (1985). *Opt. Eng.* 24: 267.

Kardar, M., Parisi, G. & Zhang, Y. -C. (1986). *Phys. Rev. Lett.* 56: 889.

Kirtley, J. R. (2009). *Supercond. Sci. Technol.* 22: 064008.

Kondo, K. & Ishibashi, A. (2006). *Jpn. J. Appl. Phys.* 45: 9137.

Kondo, K., Kaiju, H. & Ishibashi, A. (2008). *Mater. Res. Soc. Symp. Proc.* 1067: B0301.

Kondo, K., Kaiju, H. & Ishibashi, A. (2009). *J. Appl. Phys.* 105: 07D522.

Kondo, K., Kaiju, H. & Ishibashi, A. (2010). *Mater. Res. Soc. Symp. Proc.* 1198: E0701.

Kondo, K. (2010). *J. Appl. Phys.* 107: 09C709.

Kuiper, P., Kruizinga, G., Ghijsen, J., Sawatzky, G. A. & Verweij, H. (1989). *Phys. Rev. Lett.* 62: 221.

Kwon, S., Yan, X., Contreras, A. M., Liddle, J. A., Somorjai, G. A. & Bokor, J. (2005). *Nano Lett.* 5: 2557.

Lau, C. N., Stewart, D. R., Williams, R. S. & Bockrath, M. (2004). *Nano Lett.* 4: 569.

Mayadas, A. F. & Shatzkes, M. (1970). *Phys. Rev. B* 1: 1382.

Melosh, N. A., Boukai, A., Diana, F., Gerardot, B., Badolato, A., Petroff, P. M. & Heath, J. R. (2003). *Science* 300: 112.

Mendes, P. M., Flood, A. H. & Stoddart, J. F. (2005). *Appl. Phys. A* 80: 1197.

Miyazaki, T. & Tezuka, N. (1995). *J. Magn. Magn. Mater.* 139: L231.

Moodera, J. S., Kinder, L. R., Wong, T. M. & Meservey, R. (1995). *Phys. Rev. Lett.* 74: 3273.

Nacereddine, C., Layadi, A., Guittoum, A., Cherif, S. -M., Chauveau, T., Billet, D., Youssef, J. B., Bourzami, A. & Bourahli, M. -H. (2007). *Mater. Sci. Eng. B* 136: 197.

Naulleau, P. P., Anderson, C. N., Chiu, J., Denham, P., George, S., Goldberg, K. A., Goldstein, M., Hoef, B., Hudyma, R., Jones, G., Koh, C., Fontaine, B. L., Ma, A., Montgomery, W., Niakoula, D., Park, J.-o., Wallow, T. & Wurm, S. (2009). *Microelectron. Eng.* 86: 448.

Nawate, M., Shinohara, K., Honda, S. & Tanaka, H. (2004). *Trans. Mater. Res. Soc. Jpn.* 29: 1599.

Ohkawa, K., Nakano, T. & Baba, S. (2002). *J. Vac. Soc. Jpn.* 45: 134.

Ono, K., Shimada, H. & Ootuka, Y. (1997). *J. Phys. Soc. Jpn.* 66: 1261.

Palasantzas, G. & Krim, J. (1994). *Phys. Rev. Lett.* 73: 3564.

Parkin, S. S., Kaiser, C., Panchula, A., Rice, P. M., Hughes, B., Samant, M. & Yang, S.-H. (2004). *Nat. Mater.* 3: 862.

Pease, A. R., Jeppesen, J. O., Stoddart, J. F., Luo, Y., Collier, C. P. & Heath, J. R. (2001). *Acc. Chem. Res.* 36: 433.

Rahaman, M. D., Kaiju, H., Kawaguchi, N. & Ishibashi, A. (2008). *Jpn. J. Appl. Phys.* 47: 5712.

Reed, M. A., Chen, J., Rawlett, A. M., Price, D. W. & Tour, J. M. (2001). *Appl. Phys. Lett.* 78: 3735.

Rugar, D., Budakian, R., Mamin, H. J. & Chui, B. W. (2004). *Nature* 430: 329.

Sanchez, A., Davis, C. F., Liu, K. C. & Javan, A. (1978). *J. Appl. Phys.* 49: 5270.

Sawatzky, G. A. & Allen, J. W. (1984). *Phys. Rev. Lett.* 53: 2339.

Sewell, H., Chen, A., Finders, J. & Dusa, M. (2009). *Jpn. J. Appl. Phys.* 48: 06FA01.

Tarlov, M. J. (1992). *Langmuir* 8: 80.

Thompson, B. C. & Fréchet, J. M. (2008). *J. Angew. Chem. Int. Ed.* 47: 58.

Thompson, C., Palasantzas, G., Feng, Y. P., Sinha, S. K. & Krim, J. (1994). *Phys. Rev. B* 49: 4902.

Tong, W. M., Williams, R. S., Yanase, A., Segawa, Y. & Anderson, M. S. (1994). *Phys. Rev. Lett.* 72: 3374.

Troeman, A. G. P., Derking, H., Borger, B., Pleikies, J., Veldhuis, D. & Hilgenkamp. H. (2007). *Nano Lett.* 7: 2152.

Vries, J. W. C. (1987). *Thin Solid Films* 150: 209.

Wang, C. S. & Callaway, J. (1977). *Phys. Rev. B* 15: 298.

Wilke, I., Oppliger, Y., Herrmann, W. & Kneubuehl, F. K. (1994). *Appl. Phys. A* 58: 329.

Wu, W., Jung, G.-Y., Olynick, D. L., Straznicky, J., Li, Z., Li, X., Ohlberg, D. A. A., Chen, Y., Wang, S. -Y., Liddle, J. A., Tong, W. M. & Williams, R. S. (2005). *Appl. Phys. A* 80: 1173.

Yuasa, S., Nagahama, T., Fukushima, A., Suzuki, Y. & Ando, K. (2004). *Nat. Mater.* 3: 868.

12

Extremely Wetting Pattern by Photocatalytic Lithography and Its Application

Yuekun Lai[1,2], Changjian Lin[1*] and Zhong Chen[2*]
*[1]State Key Laboratory for Physical Chemistry of Solid Surfaces and
College of Chemistry and Chemical Engineering, Xiamen University*
[2]School of Materials Science and Engineering, Nanyang Technological University
[1]China
[2]Singapore

1. Introduction

Wettability is an important property governed by not only chemical composition, but also geometrical structure as well (Ichimura et al., 2000; Lai et al., 2009a; Wang et al., 1997). Super-wetting and antiwetting interfaces, such as superhydrophilic and superhydrophobic surfaces with special liquid-solid adhesion have recently attracted worldwide attention. Superhydrophilicity and superhydrophobicity are defined based on the conventional water contact angle experiment. If the contact angle is smaller than 5°, the surface is said to be superhydrophilic. Superhydrophobic refers to surface with contact angle greater than 150°. Such two extremely cases have attracted much interest due to their importance in both theoretical research and practical application (Lafuma & Quéré, 2003; Liu et al., 2010; Gao & Jiang, 2004).

In recent years, patterned thin films have received considerable attentions due to their interesting properties for a range of applications, such as optoelectronic devices, magnetic storage media, gas sensors, and fluidic systems. Compared to the conventional thin film technology, such as physical vapor deposition (Li et al., 2006; Zhang & Kalyanaraman, 2004), chemical vapor deposition (Jeon et al., 1996; Slocik et al., 2006) and sputtering (Rusponi et al., 1999), solution-based deposition method is becoming popular for the fabrication of patterning films due to the low temperature process under ambient environment, less energy and time consumption, and easier control of the experimental parameters (Lai et al., 2010a, Liu et al., 2007; Yoshimura & Gallage, 2008). Although traditional photolithographic technique is excellent for preparing sub-micrometer or even only sub-100-nanometer pattern (Cui & Veres, 2007; Li et al., 2009), it is a complex multi-step process (wafer cleaning; barrier layer formation; photoresist coating; soft-baking; mask alignment; exposure and development; and hard-baking) and needs to remove part of the film and all the photoresist used. Direct and selective assembly of nanostructured materials from precursors paves a new avenue for the fabrication of electronic optical microdevices.

Wetting micropatterns with different physical or chemical properties, without the need for ultra-precise positioning, have frequently been acted as templates for fabricating various functional materials in a large scale. The great difference in contact angle of the two extreme cases provides a potentially powerful and economical platform to directly and precisely construct patterned nanostructures in aqueous solution. In general, wetting micropatterns with low contact angle contrast (≤120°) on smooth substrates can be formed by photolithography (Falconnet et al., 2004; Kobayashi et al., 2011), microcontact lithography (Csucs et al., 2003; Kumar et al., 1992), colloidal patterning (Michel et al., 2002; Bhawalkar et al., 2010), electron beam lithography (Wang & Lieberman, 2003; Zhang et al., 2007), nanoimprint lithography (Jiao et al., 2005; Zhang et al., 2006), dip-pen nanolithography (Huang et al., 2010a; Lee et al., 2006; Xu & Liu, 1997), and so on. Among these methods, photocatalytic lithography employing semiconductors to photocatalytic decompose of organic monolayer is one of the most practical techniques because it able to accurately transfer an entire photomask pattern to a target substrate at a single exposure time under environmental condition (Bearinger et al., 2009; Lee & Sung, 2004; Nakata et al., 2010; Tatsuma et al., 2002; Wang et al., 2011). Moreover, it can greatly reduce the photoresist waste. The resolution of the patterning is greatly dependant on the mask alignment and light source exposure. Under optimal condition, a resolution of micrometer- or submicrometer-scale pattern of alkylsiloxane self-assembled monolayers can be achieved with UV light projection irradiation. Once patterned on the surface, organic monolayer has been applied in various ways to restrict corrosion or induce nanostructures growth. Firstly, patterned layer itself may serve as etching mask to protect the substrate to generate pattern with certain thickness/aspect ratio. Secondly, patterned organic layer may be employed as barrier to inhibit the liquid phase deposition of nanostructures to generate functional composite pattern with diverse shape and density. So far, only a few reports have been available on the fabrication and application of superhydrophilic–superhydrophobic patterning by photocatalytic lithography under ambient conditions (Lai et al., 2008a; Nishimoto et al., 2009; Zhang et al., 2007).

Upon UV irradiation, the electron-hole pairs in semiconductor TiO_2 can be generated and migrated to its surface, where the hole reacts with OH- or adsorbed water to produce highly reactive hydroxyl radicals (Zhao et al., 1998). These hydroxyl radicals can further oxidize and decompose most organic compounds. Recently, we found that the pollutant solution can be rapidly decomposed on a nanotube array TiO_2 film with UV irradiation (Lai et al., 2006, 2010b; Zhuang et al., 2007). Considering its effectiveness for the photocatalytic decomposition of organic compounds, the photocatalysis of such TiO_2 nanotube film can be a promising way to decompose the low energy hydrophobic fluoroalkyl chains. So it is possible to achieve a conversion from superhydrophobicity to superhydrophilicity due to the amplification effect of the rough aligned nanotube structure. By using a patterned photomask to control the site-selective decomposition by UV light, that is photocatalytic lithography, superhydrophilic cells can be accurately transferred to a target substrate at a single exposure time under environmental condition. Therefore, these two types of extreme wettability coexist on the surface directly to make up of superhydrophilic-superhydrophobic pattern.

In this chapter, we firstly discuss the wettability on TiO_2 nanostructure film by electrochemical anodization. Secondly, we demonstrate using a novel synthetic process to prepare wetting pattern with a high contrast (superhydrophilic–superhydrophobic) on TiO_2 nanotube structured film by a combination of SAM technique and photocatalytic lithography. The resultant micropattern has been characterized with scanning electron microscopy, optical microscopy, electron probe microanalyzer and X-ray photoelectron spectroscopy. Finally, we focus on the technological details and potential future application of wetting template to induce and direct the assembly of functional nanostructure to form uniform micropatterns. For example, the patterning, biomedical and sensing application of the wetting template and functional composite nanostructure pattern (TiO_2, ZnO, OCP and CdS) (Lai et al., 2009b, 2010a-d).

2. Wettability on TiO₂ nanostructures by electrochemical anodization

Wettability of solid surfaces is a very important property of solid surface. Surfaces with extreme wetting properties, e.g. superhydrophilic and superhydrophobic, can be prepared by introducing certain rough structures on the originally "common" hydrophilic and hydrophobic surfaces. Various ways of preparing TiO_2 semiconductor films on the different solid substrates have been developed, including sol-gel technique (Shen et al., 2005), sputtering (Takeda et al., 2001), chemical vapor deposition (Rausch & Burte, 1993), liquid phase deposition (Katagiri et al., 2007), hydrothermal (Yun et al., 2008), and electrochemical anodizing. Among them, the electrochemical anodizing is verified to be a convenient technique for fabricating nanostructured TiO_2 films on titanium substrates (Lai et al., 2004, 2008b, 2009c; Gong et al., 2003). Moreover, the conductive titanium support substrate can be an advantage for fabricating functional material composites through electrochemical depositions to further improve their photoelectrochemical activities.

Figure 1a shows a typical FESEM image of the titanium substrate before electrochemical anodization. The surface of the substrate was relatively smooth, with features of parallel polished ridges and grooves at the micron scale (Lai et al., 2010a). Figure 1b shows the top view SEM image of the typical TiO_2 nanotube array film by anodizing under 20 V for 20 min. After anodization, shallow cavities as large as several micrometers in diameter were present on the surface of the sample. This is probably due to the anisotropic oxidation of the underlying Ti grains (Crawford & Chawla, 2009; Yasuda et al., 2007). From the high magnification image (Fig. 1c), it can be seen that vertically aligned TiO_2 nanotubes with inner diameter of approximately 80 nm covered the entire surface including the shallow polygonal micropits. The side view image shows that the self-assembled layers of the TiO_2 nanotubes were open at the top and closed at the bottom with thickness about 350 nm (inset of Fig. 1c). Water droplet can quickly spread and wet the as-grown vertically aligned TiO_2 nanostructure film due to capillary effect caused by the rough porous structure, indicating such TiO_2 structure film by electrochemical anodizing is superhydrophilic. A more hydrophobic behaviour, on the other hand, was obtained after coating the TiO_2 film with fluoroalkyl silane. The inset of Fig. 1b shows the intrinsic contact angle (CA) on the as-prepared vertically aligned TiO_2 nanotube surface and its corresponding 1H,1H,2H,2H-perfluorooctyltriethoxysilane (PTES, Degussa Co., Ltd.) modified surface is nearly 0° (superhydrophilic) and 156° (superhydrophobic), respectively. However, the CA for the "flat" TiO_2 surface and its corresponding PTES modified sample is about 46° (hydrophilic) and 115° (hydrophobic), respectively. From these results, we know the top surface of the vertically aligned nanotubes has an amplification effect to make hydrophilic and hydrophobic surfaces become superhydrophilic and superhydrophobic, respectively. After UV irradiation for 30 min, the water CA on the TiO_2 nanotube film and "flat" TiO_2 film decreased to 0° and 26°, as a consequence of the photocatalytic activity of TiO_2 films (Balaur et al., 2005; Lai et al., 2010a). Moreover, the sample showed hydrophobic character once again when it was treated with PTES. Therefore the surface can be reversibly switched between superhydrophobic and superhydrophilic by alternating SAM and UV photocatalysis on the rough TiO_2 nanotube arrays (shown in Fig. 1d). Compared with the large wettability contrast on this type of rough surface (larger than 150°), the wettability of a "flat" TiO_2 film can only be reversibly changed within the small range between 26° and 115°.

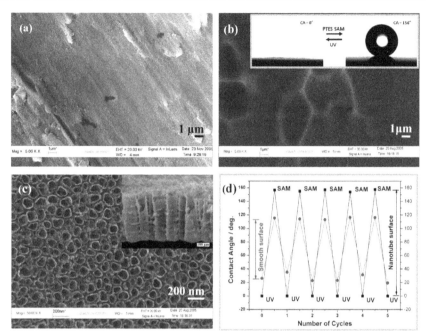

Fig. 1. SEM images of the mechanically polished and cleaned titanium substrate (a), low magnification of a nanotube structured TiO$_2$ film (b), and a higher magnification of the TiO$_2$ nanotube array film (c). Reversible surface wettability on a "flat" TiO$_2$ film and rough nanotube TiO$_2$ film by alternating SAM and UV photocatalysis (d). The inset of (b) shows the shape of a water drop on the PTES-modified and UV-irradiated TiO$_2$ nanotube array film. The inset of (c) shows the side view of a TiO$_2$ nanotube array film.

Recently, we designed three types of superhydrophobic nanostructure models consisting of a nanopore array (NPA), a nanotube array (NTA), and a nanovesuvianite structure (NVS) apply a facile electrochemical process (Figure 2) (Lai et al., 2009a). Based on basic principles of roughness-enhanced hydrophobicity and capillary-induced adhesion, these different porous structures were expected to create interfaces with decreasing adhesive forces. The surface adhesive forces could be effectively tuned by solid–liquid contact ways at the nanoscale and air-pocket ratio in open and sealed systems. The magnitude of the adhesive force of a droplet for a superhydrophobic surface descends in the order "area-contact" > "line-contact" > "point-contact". A continuous three-phase (solid–air–liquid) contact line (TCL) generates serious CA hysteresis and surface adhesion, while a discrete TCL is energetically advantageous to drive a droplet off a superhydrophobic surface, showing lower surface adhesion. Therefore, the water droplet behavior on these superhydrophobic surfaces could be greatly changed from pinning to sliding by adjusting the solid-liquid contact way.

Capillary adhesive force plays a dominant role in imparting adhesive behavior on NPA and NTA nanostructures will sealed cells, while the open NVS nanostructure, which had extremely low adhesion capacity for water, acted solely by van der Waals attraction between water and PTES molecules. A possible explanation is as follows. As the droplet gradually retracted from the sample surface, the meniscus on each nanotube nozzle would be changed from concave to convex (Figure 3a). This could result in an increased volume of air sealed in each nanotube by the liquid/air interface. According to Boyle's law (West, 1999), there is an inverse relationship between the volume (V) and pressure (P) for an ideal gas under the conditions of constant temperature and quality. Therefore, this expansion of air would result in the formation of a negative pressure (ΔP). In this case, the volume of air sealed in the nanotubes was varied by their depths, so longer tubes would be expected to have lower air-expansion ratios ($\Delta V/V$), thus lesser negative pressures. For a fixed nanotube diameter, a longer nanotube would therefore require a smaller pulling-off force, and the total surface adhesive force would be smaller.

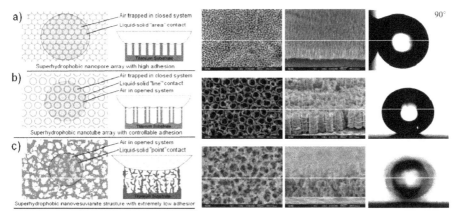

Fig. 2. Schematic models, SEM images and corresponding water behavior on three types of superhydrophobic porous-nanostructure with water adhesive forces ranging from high to low. (a) Superhydrophobic NPA with high adhesion. (b) Superhydrophobic NTA with controllable adhesion. (c) Superhydrophobic NVS with extremely low adhesion.

Figure 3b shows the curves of the water CAs and adhesive forces with respect to the diameter of individual nanotube. When the nanotube diameter decreases, the force drastically increased, while the CA slightly decreased. When the diameter was tuned from 78 nm down to 38 nm, the surface adhesive force of the superhydrophobic NTA film increased 2.06 times, while the decline in magnitude of the water CA was not more than 2%, showing that the negative pressure caused by the volume change of air sealed in the nanotubes could effectively tune the surface adhesive force. The NTA structures in this study had variable length, with values of (0.35 ± 0.04) µm, (0.76 ± 0.05) µm, (1.18 ± 0.07) µm, while their diameter (~80 nm) was fixed. Figure 3c shows the curves for water CAs and adhesive forces obtained with PTES-modified NTA-nanostructure surfaces differing in nanotube length. With lengths extending from 0.35 µm to 0.76 µm and 1.18 µm, the CA change was very small, not more than 2%, which could be due to the minor variation in nanotube diameter. However, the adhesive force linearly decreased from 21.5 µN down to 16.7 µN and 12.2 µN for the above increases in length, respectively. It was evident that the water adhesive force of the superhydrophobic NTA-nanostructure surfaces could be tuned by varying the diameters and also lengths of the nanotubes. These findings are valuable to deepen insight into the roles of nanostructures in tailoring surface water-repellent and adhesive properties for exploring new applications.

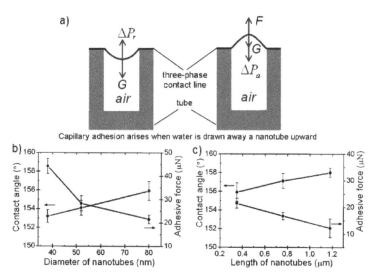

Fig. 3. (a) Capillary adhesion arises when a water droplet sitting on the tube nozzle is gradually drawn upwards because the convex air/liquid interface produces an inward pressure ΔP. (b,c) The curves of water contact angles and adhesive force on the superhydrophobic NTA nanostructures with respect to the diameter and length of nanotubes.

3. Wetting pattern by photocatalytic lithography

A novel approach for constructing superhydrophilic-superhydrophobic micropattern on the nanotube structured TiO_2 films has developed by using photocatalytic lithography (Figure 4a) (Lai et al., 2008a). At the first step, the as-prepared amorphous TiO_2 nanotubes by electrochemical anodizing of titanium sheets were calcinated at 450°C to form anatase phase, then treated with a methanolic solution of hydrolyzed 1 wt% PTES for 1 h and subsequently heated at 140°C for 1 h, and at the second step, the superhydrophobic film is selectively exposed to UV light through a copper grid (photomask) to photocatalytically cleave the fluoroalkyl chain. It is noteworthy, from the characterization of chemical composition before and after UV irradiation by X-ray photoelectron spectroscopy, that the intensities of the F1s and FKLL are decreased greatly and those of the Ti2p and O1s are increased after exposing the PTES modified surface to UV light for 20 min (Fig. 4b). From the inset high-resolution spectra (Fig. 4c), the peaks of -CF_2 (at 291.8 eV) and -CF_3 (at 294.1 eV) are obviously vanished after UV light irradiation, while the strength of silicon peaks in the XPS spectra remains unchanged but shifts from 102.8 to 103.3 eV, suggesting that Si-O-Si networks have already formed due to UV irradiation. According to these results, we believe that the hydrophobic fluoroalkyl chains have been completely decomposed and removed by the photocatalytic reactions at TiO_2 nanotube films. Similarly, a serial of fluoroalkyl silane monolayer pattern (e.g. heptadecafluorodecyltrimethoxysilane, octadecyltriethoxysilane, and methyltriethoxysilane) can be successfully fabricated in our case. Although various monolayer patterns can be prepared with a resolution about micro-scale or submicro-scale under optimal condition, we will focus on the application of the PTES micro-pattern with a general TEM copper grid as a photomask by photocatalytic lighography.

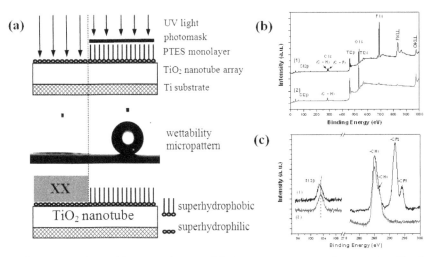

Fig. 4. Schematic outline of the procedures to fabricate nanostructured patterning film by electrochemical deposition based on superhydrophilic-superhydrophobic micropattern (a). Survey-scan X-ray photoelectron spectra of the PTES modified nanotube TiO_2 films before (1) and after (2) 20 min UV irradiation (b). The high-resolution spectra of Si2p and C1s regions (c).

Figure 5 shows the optical micrograph of the as-obtained superhydrophilic–superhydrophobic pattern by focusing on the droplet within the superhydrophilic regions. A uniform pattern is formed due to the site-selective wetting by water droplets within the superhydrophilic regions (Fig. 5a). A light dot array (inset of Fig. 5a) is seen when focusing on the top of the droplets, indicating that the confined droplet has a hemispherical dome. To further verify the resulting micropatterns with an extreme wettability contrast, fluorescein sodium was used as a probe to label the surface of the films. Figure 5b shows the fluorescent micrograph of the resultant superhydrophilic–superhydrophobic micropatterns on the TiO_2 nanotube array films. As shown, geometrically identical square superhydrophilic regions and dark superhydrophobic regions transferred well from the photomask to form a well-defined pattern. The UV-irradiated regions become superhydrophilic owing to the photocatalytic cleavage of the PTES molecule and the enhanced roughness of the nanotube structure, while the non-irradiated parts remain superhydrophobic without any change. Because the difference in the water CA between the irradiated and non-irradiated regions is larger than 150°, the liquid containing the fluorescent probe selectively appears only on the uniform superhydrophilic grids and not on the neighboring superhydrophobic regions. Therefore, a clear, well-defined fluorescent pattern in line with the dimensions of the Cu grid can be obtained. These results indicate that the micropatterned template composed of superhydrophilic and superhydrophobic regions was fabricated successfully.

The UV irradiation times had a great effect on the quality of the resulting pattern. For example, it cannot exhibit a sufficient wettability contrast between the irradiated and non-irradiated regions to form a well-defined pattern within 5 min. This is attributed to the hydrophobic fluoroalkyl chain in the PTES molecule that was not efficiently cleaved under a short-time UV irradiation. However, with a long-time UV irradiation (i.e., 60 min), the adjacent PTES molecule covered by the Cu grid can be remotely oxidized by a TiO_2 nanotube photocatalyst or the diffusion, scattering, and diffraction of the incident light (Haick & Paz, 2001; Kubo et al., 2004). Therefore, to obtain a higher pattern resolution, the optimized UV irradiated time in our case was controlled in the range of 10-30 min.

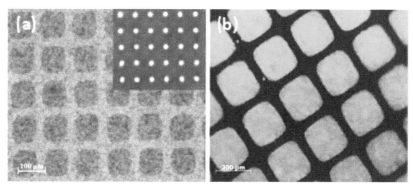

Fig. 5. (a) Optical micrograph of the as-obtained superhydrophilic-superhydrophobic pattern by focusing on the water droplet within the superhydrophilic regions. (b) Fluorescence microscopy image of the fluorescein probes on the as-prepared superhydrophilic-superhydrophobic micropattern.

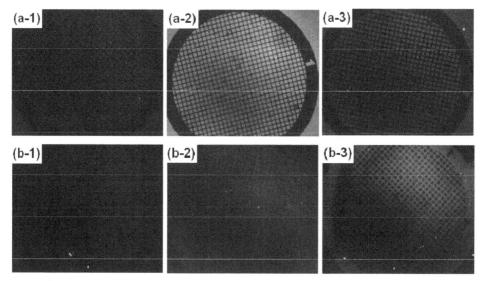

Fig. 6. Fluorescence microscope images generated by blue light excitation on the different micropatterned templates with the adsorption of FITC-BSA, (a-1, a-2, a-3) the PTES template with a pH value of 7.5, 4.5 and 2.5, respectively; (b-1, b-2, b-3) the PTES-APTS template with a pH value of 7.5, 4.5 and 2.5, respectively.

Based on the molecular self-assembly and photocatalytic lithography techniques, micropatterned templates of PTES or PTES-APTS(3-aminopropyltriethoxysilanes) with different wettabilities were fabricated on the titania film. The adsorption behavior of bovine serum albumin (BSA) on the above two templates was investigated using fluorescent labeling (FITC) in buffer solution with different pH values. The results showed that, for the PTES template with great wettability differences, BSA would preferentially adsorb on the superhydrophilic regions. For the APTS-PTES template with smaller differences in wettability, competitive adsorption phenomenon on the super-hydrophobic regions was found due to the hydrophobic interaction force between the albumin and the surface. As the pH value decreased to 2.5, the phenomenon of competitive adsorption was prominent with the albumin adsorbed in the super-hydrophobic areas. The adsorption feature of the albumin may be closely related to the wettability and surface energy of the materials. This technique has promising applications in bio-compatible coatings where drugs could be encapsulated in specific areas of the coating using simple microfabrication methods.

4. Application of wetting pattern

Uniform self-assembly of functional inorganic nanomaterials is a fundamental challenge. Nature adopts a superior approach in biomineralization, where "matrix" macromolecules induce nucleation of inorganic crystals at specific locations with controlled size and morphology, and sometimes even with defined growth orientation. We apply the biomimetic principles derived from liquid phase processes to the assembly of nanoscale functional materials into microscale systems. We carefully control surface wettability to promote etching or heterogeneous nucleation at designated superhydrophilic regions while completely suppress these processes elsewhere (superhydrophobic regions), therefore enable the controlled top-down or bottom-up assembly of inorganic nanomaterials directly from solution. Following this principle, arrays of crystalline TiO_2 nanotube, ZnO nanorods, CdS semiconductor materials and octacalcium phosphate (OCP) biomaterials were nucleated and assembled directly from solution onto Ti substrates at the desired precise locations and then fabricated into arrays of photodetector or matrix devices for large-area microelectronic applications. This strategy of micropatterned nanocomposites will be helpful to develop various micropatterned functional nanostructured materials.

4.1 Template for preparing functional pattern

Figure 7 shows an optical microscopy image of the TiO_2 nanotube micropattern produced using a grid micropattern with different wet etching times (Lai et al., 2009b). A patterning with a clear outline was formed in a short time for 30 s (Fig. 7a). With an increase in the wet etching time (Fig. 7b and c), identical micropatterns with higher aspect ratios can be fabricated. When the etching was prolonged to 240 s (Fig. 7d), the size of the grids increased slightly, indicating that the PTES-SAM layer at the edge of the superhydrophobic lines is more easily etched as compared to the inner superhydrophobic area. This is due to the loose and disordered SAMs resulting from the scattered UV light photocatalytic degradation and the transfer of the active hydroxyl radicals at the edge of the grids. Moreover, the isotropic etching of the Ti substrate underneath the boundary leads to the collapse of the upper nanotube array structures. Therefore, the pattern can be obtained with a clear boundary in a short time after the wet etching in the aqueous solution, and the depth of the etching can be controlled simply by adjusting the etching time.

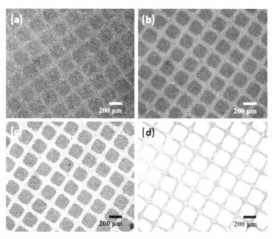

Fig. 7. Optical micrograph images of the time-resolved evolution process of the resulting vertical aligned TiO_2 nanotube micropattern on a superhydrophilic-superhydrophobic template by wet chemical etching in 0.1 wt% HF solution: (a) 30, (b) 60, (c) 120, and (d) 240 s.

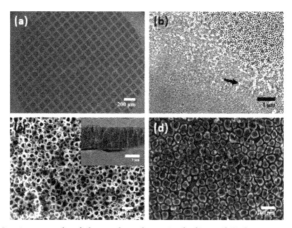

Fig. 8. (a) The SEM micrograph of the ordered vertical aligned TiO_2 nanotube array pattern by vapor-condense etching for 10 min. Magnified images of the corresponding (b) superhydrophilic-superhydrophobic boundary, (c) superhydrophobic area, and (d) superhydrophilic area. The inset figure shows the corresponding cross-sectional image.

The micropattern with a higher aspect-ratio can also be fabricated by a developed vapor etching technique. As water evaporated, the vapor containing HF dewetted the superhydrophobic lines while condensing on the superhydrophilic grids to etch the vertical aligned TiO_2 nanotube layer. Figure 8a and b shows the as-prepared TiO_2 micropattern with a clear boundary by water vapor etching for 10 min. This resulted from the selective condensing of the vapor from the water solution containing 5 wt% HF in the superhydrophilic regions. Some collapsed residue (indicated by the black arrow) covers the boundary due to the rapid etching of the bottom nanotubes and the resultant bubbles;

however, they can be easily removed by sonication. The vertical aligned TiO_2 nanotube film with a length of about 1.53 µm shows no obvious change within the superhydrophobic regions (Fig. 8c), while the TiO_2 nanotubes in the superhydrophilic regions are etched completely for 20 min (Fig. 8d), that is to say, a TiO_2 micropattern with a high aspect ratio of ~20 can be obtained. This technique is particularly attractive in generating large-area functional nanostructure patterns in a high throughput fashion with a high aspect-ratio.

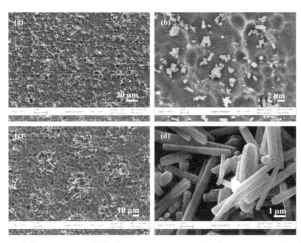

Fig. 9. SEM images of the patterned ZnO nanostructured micropatterns by liquid phase deposition at 90°C for different times: (a, b) 30 min; (c, d) 90 min.

Figure 9 shows representative top-view FESEM images of the ZnO/TiO_2 micropatterns by liquid-phase deposition after different times. After growth for 30 min (Fig. 9a,b), the nucleation and growth of ZnO crystals with various morphologies and sizes (nanoparticles and nanorods) are sparsely dispersed within the predefined superhydrophilic regions. In the superhydrophobic regions, the nanotube structure is retained with its original morphology due to the indirect contact with the solution and the effective protection by the PTES monolayer. Upon further increase in the deposition time to 90 min (Fig. 9c,d), it was observed that ZnO nanorods were the predominant structural features. The average diameter of the grown ZnO nanorods increases greatly to about 400–800 nm in diameter and 3-5 µm in length, which may be due to lower nucleation rate and higher growing space for ZnO nanorods. The superhydrophilic microdots are almost covered with randomly lying ZnO nanorods. Moreover, there are different growth orientations and some connections into adjacent nanorods. It is of interest to note that a two dimensional (2D) pattern with smaller density of randomly packed ZnO nanorods, instead of the 3D pattern consisted of well-aligned vertical nanorods, which were confined and grown within the superhydrophilic regions.

For practical application of thin film devices, the position and orientation-control of ZnO nanorods are very important because they directly relate to their physical and chemical performances (Koumoto et al., 2008; Masuda et al., 2006). In order to precisely control the spatial orientation of the ZnO nanostructures, we developed a new technique which is able to make the ZnO nanorods grow along the vertically aligned titania nanotubes rather by disordered deposition on the titania nanotube array surface. In this technique, resistance discrepancy was adopted to make the entrance of the tubes more conductive than the bottom of the tubes, to induce the epitaxial growth with spatial organization of uniform

ZnO nanorods along the direction of the nanotubes. The quasi-perpendicular ZnO nanorods nucleate and grow uniformly and selectively throughout the superhydrophilic regions of the TiO_2 nanotube surface by electric field assisted deposition at 90°C for 3 min, while no nanorods are observed in the superhydrophobic regions (Fig. 10a,b) (Lai et al., 2010a). The EDS spectra also reveal that the presence of Zn, Ti and O elements on the superhydrophilic regions, while the elemental components in the superhydrophobic areas are only Ti and O. The inset of Fig. 10b shows the hexagonal end facet of a vertically aligned ZnO nanorod with a diameter about 100-150 nm growing on top of the TiO_2 nanotube array surface. Therefore, the density, size and orientation of ZnO nanorods are very sensitive to the presence of electric fields. A 3D AFM profile image (Fig. 10c) shows that the microscopic structure of the ZnO crystal deposition consisted in dense column arrays, which are induced and directed by the wettability template. The thickness of vertical ZnO nanorod film is in the range of 800-900 nm. Furthermore, the three dimensional confocal microscopy image (Fig. 10d) also shows that the growth of the ZnO nanorod pattern is identical with the superhydrophilic/superhydrophobic template.

On the basis of the versatile superhydrophilic-superhydrophobic template, we can successfully control the growth of ZnO nanostructures in the superhydrophilic regions under mild reaction conditions and in the absence of seed and noble metal catalyst. In the superhydrophobic regions, the growth is suppressed. This special template can be utilized to generate different nanostructured ZnO patterns with clearly defined edges. Hence, it is expected that this novel micropatterned technique based on the superhydrophilic-superhydrophobic template will become a powerful tool for fabricating various types of micropatterned nanomaterials and devices.

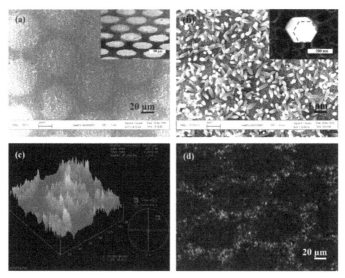

Fig. 10. (a,b) Typical SEM images of the vertically aligned ZnO nanorods selectively grown on superhydrophilic patterning regions by the developed electric field assisted deposition technique at 90°C for 3 min. The inset in (a) shows the side view SEM image of the corresponding ZnO nanorod micropattern. The inset in (b) shows the higher magnified SEM image of a ZnO nanorod with hexagonal end facet. (c) 3D AFM image of the ZnO/TiO_2 micropattern. (d) Confocal microscopy image of the perpendicular ZnO nanorod array.

4.2 Biomedical arrays

The typical SEM image of TiO$_2$ nanotube array surface before and after the deposition of OCP film by electrochemical technique for 5 min is shown in Fig. 11a,b. The superhydrophobic-superhydrophilic micropatterned TiO$_2$ was used as a micro-template to selectively deposit nano-OCP crystals on the superhydrophilic regions by an electrochemical deposition to form a special micropatterned nano-OCP. The deposition electrolyte was consisted of 0.042 mol/L Ca(NO$_3$)$_2$ and 0.025 mol/L NH$_4$H$_2$PO$_4$. The pH value was adjusted to approximately 4.2 with 0.05 mol/L NaOH solution. The precipitation was carried out galvanostatically at a cathodic current of 0.5 mA cm^{-2} under 67.5°C for a certain time (Wang et al., 2008). It can see that quasi-perpendicular ribbon-like crystals of several hundred nanometers in width are uniformly grown on the TiO$_2$ nanotube array surface. Fig. 11c shows a typical fluorescence microscope of the superhydrophilic-superhydrophobic micropattern on TiO$_2$ nanotube surface. As can be seen, the green dot patterns are clearly imaged through the fluorescence contrast between the UV-irradiated superhydrophilic and photomasked superhydrophobic regions. The photoirradiated dot exhibiting a uniformly stronger fluorescence against the surrounding dark background is due to the highly affinity to solution resulting in the absorption of fluorescent probes into the irradiated nanotube array films. Therefore, a clear well-defined fluorescence pattern based on the superhydrophobic-superhydrophilic pattern is obtained. Fig. 11d displays the identical patterning of OCP biomaterials deposited on the superhydrophilic-superhydrophobic patterns on TiO$_2$ nanotube array surface. It is obvious that the size of the white OCP dots is equal to that of the superhydrophilic area on template, indicating the deposited regions were only located within the superhydrophilic dots where photocatalytic degradation of PTES-SAMs was performed.

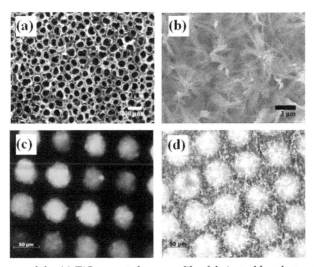

Fig. 11. SEM images of the (a) TiO$_2$ nanotube array film fabricated by electrochemical anodization; (b) OCP nanostructure layer on TiO$_2$ nanotube array film by electrochemical deposition for 5 min. Optical fluorescence pattern of the superhydrophilic-superhydrophobic template (c) and patterned OCP thin films selectively deposited in pre-defined superhydrophilic regions by electrochemical deposition for 5 min (d).

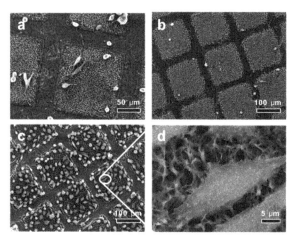

Fig. 12. SEM micrographs of MG-63 cells cultured on the patterned OCP coatings with different deposition time for 6 h, (a) 1 min; (b) 3 min; (c) 5 min; (d) higher magnification.

The *in-vitro* MG-63 cell tests were used to study the biological performance of the as-obtained OCP micropatterns (Huang et al., 2010b). The results showed that MG-63 cells were found preferentially attached on the superhydrophilic regions with OCP thin films, while the superhydrophobic regions with the PTES monolayers can effectively prevented the adhesion of cells on the surface, indicating that the cells had the selective adhesion action on the tiny units of OCP films. Moreover, the cells adhered on the OCP film deposited for a longer period (5 min) are more active to spread on the OCP nanobelt covering surface. It is promising for developing a new cell chip for high throughput evaluation of the cell behaviors.

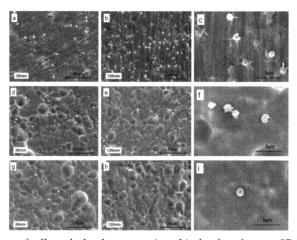

Fig. 13. SEM images of adhered platelets on various kinds of surfaces at 37°C for different periods. (a-c) mechanically polished and cleaned Ti substrate; (d-f) superhydrophilic surface; (g-i) superhydrophobic surface; (a,d,g) 30 min; (b,e,h) 120 min; (c,f,i) magnified images of the corresponding images of (b,e,h).

The *in vitro* experimental indicated that the superhydrophobic nanotube TiO_2 layers exhibit a remarkable resistance to platelets attachment (Fig. 13) (Yang et al., 2010). It is indicated that the superhydrophobic nanotube TiO_2 layers exhibit a remarkable resistance to platelets attachment. As shown in Fig. 13a,d,g, abundant platelets adhere to both the plain Ti surface and the superhydrophilic TiO_2 nanotube layers after 30 min incubation. Comparatively, after 120 min incubation, a large number of platelets adhered and spread out on both the plain Ti surface (77 ± 7.4 per 5000 μm^2, Fig. 13b,c) and the superhydrophilic surface which was obtained by exposing the TiO_2 nanotubes under a UV irradiation (22 ± 1.5 per 5000 μm^2, Fig. 13e,f), only very few of platelets (1 ± 0.8 per 5000 μm^2) was found to adhere on the superhydrophobic TiO_2 nanotube layers (Fig. 13h). Moreover, even though some platelets were occasionally seen attached on the superhydrophobic surface, they looked smooth without any growth of pseudopods (Fig. 13i), implying that the platelets adhered on the superhydrophobic TiO_2 nanotube surface remain inactive and hardly grow and spread out for a long period. The quantities and morphologies of adhered platelets and their corresponding interactions on the different samples are illustrated in Figure 14. Therefore, the construction of superhydrophobic surface on biomedical implants could pave a way to improve the blood compatibility of the biomedical devices and implants.

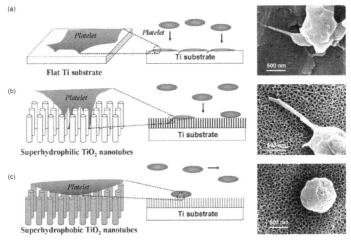

Fig. 14. Schematic illustration of the quantity and morphology of platelet and corresponding interactions on the three kinds of surfaces. (a) Plain Ti substrate; (b) Superhydrophilic TiO_2 nanotubes; and (c) Superhydrophobic TiO_2 nanotubes.

4.3 Sensing devices

Based on photocatalytic lithography, we demonstrate a facile, rapid and practical approach to fabricate Ag nanoparticle (NP) patterns on TiO_2 films by means of pulse-current electrodeposition technique (Huang et al., 2011). The size and density of as-deposited Ag NPs can be controlled by changing deposition charge density. Moreover, the resultant patterned Ag NP films exhibited particle size-as well as density-dependent UV-vis absorption and SERS enhancement effect. It was found that the patterned Ag NP films produced under the deposition charge density of 2.0 C cm^{-2} exhibited the intense UV-vis and Raman peaks. Two dimensional surface enhanced Raman scattering (SERS) mapping of

Rhodamine 6G (R6G) on the patterned Ag NP films demonstrated a high throughput localized molecular adsorption and micropatterned SERS effect.

Furthermore, the elemental distributions of the as-prepared Ag NPs arrays were also observed by electron probe microanalyzer, which are shown in Fig. 15a-c. Figure 15a shows the Ag element distribution map. As shown in the map, the green dot patterns are clearly images obtained through element concentration contrast between the UV-irradiated superhydrophilic and photomasked superhydrophobic regions. The green dots exhibiting a uniform Ag concentration against the surrounding black regions indicate that Ag NPs are uniformly deposited and confined to the superhydrophilic regions. Figure 15b,c shows the element distribution maps of Ti and O, respectively, which are also in line with the dimensions of the photomask. The blue superhydrophilic regions (dot patterns) show lower Ti and O concentrations due to the preferential deposition of Ag NPs in the superhydrophilic areas, while the yellow and red superhydrophobic regions have higher Ti and O concentration. The corresponding line-scan signal intensity profiles of Ag, O, and Ti elements across the dot pattern (red line direction). The intense Ag signals in the superhydrophilic regions as well as the Ti and O signals in the superhydrophobic regions suggest that Ag NPs are deposited only in the dot areas and that the other regions are the exposed TiO_2 nanotube films. The consistent intensity of the Ag signals indicates that Ag NPs are uniformly deposited on the superhydrophilic regions.

In addition, a two-dimensional point-by-point SERS mapping of the patterned Ag NP film whose deposited charge density is 2.0 C cm^{-2} was obtained using R6G as the probe molecule. Figure 15d,e show the optical image and the corresponding SERS mapping image of the patterned Ag NP film. The mapping area was approximately 140 × 100 µm^2 and the data acquisition time was 1 s. A signal to baseline from 594.0 to 623.4 cm^{-1} was chosen for the acquisition of the SERS mapping. The bright and dark areas respectively represent higher and lower intensity of the SERS signal. It is clear that the geometrically identical gray superhydrophilic areas (circle) with a strong SERS activity and the dark superhydrophobic areas without any SERS activity form a high-resolution SERS intensity distribution map. As can be seen from the SERS mapping results, most SERS peak area is in a very narrow intensity window as shown by the contrast in color codes. Furthermore, the SERS peak area is uniform over the superhydrophilic region with several high intensity spots represented by white color codes. On comparing the SERS mapping with the SEM detection, it is reasonable to conclude that the homogeneous SERS signal in the circle areas reflects the uniform dispersion of Ag NPs on the superhydrophilic areas. The high-resolution SERS intensity distribution and micropatterned SERS effect of the Ag NP film might make it potentially useful in high-throughput molecule detection and bio-recognition.

To gain insight into the dependence of SERS enhancement on the size and density of Ag NPs, the SERS spectra of R6G absorbed on the different patterned Ag NP films were detected, which are shown in Fig. 15f. Because of the small particle size and low density, which are not the optimum size and distribution for SERS, the signal enhancement is rather weak below the charge density of 0.5 C cm^{-2}. The enhancement behavior of the substrate, however, is obviously improved under the charge density of 1.0 C cm^{-2}. In particular, the patterned Ag NP film prepared under a charge density of 2.0 C cm^{-2} exhibits the highest intensity, which is attributed to large size and high density of Ag NPs, as shown by the SEM results. On increasing the charge density to 2.5 C cm^{-2}, the signal becomes weaker. The size-correlated enhancement may be explained by the EM mechanism (Zeng et al., 2008). The intensity of the SERS signal might also be controlled by the NPs density, which changes the interparticle spacing as well as the hot spots among the NPs (Felidj et al., 1999; Lu et al., 2005). The average surface enhancement factor for pyridine on the Ag NP film with a charge density of 2.0 C cm^{-2} was calculated to be 1.3×10^5.

Fig. 16a shows the typical SEM micrographs of the CdS nanosphere micropatterns after 3 min deposition on the superhydrophobic-superhydrophilic template of TiO$_2$ nanotube films (Lai et al., 2010d). The bright rectangular areas corresponded to the deposition of CdS nanosphere crystals on the superhydrophilic regions. The boundary between the CdS pattern and the surrounding superhydrophobic regions is clearly visible at a higher magnification (inset). The dispersed CdS nanosphere crystals grew on the top of TiO$_2$ nanotube arrays within the rectangular superhydrophilic region (Fig. 16c). Most of the crystals were less than 90 nm in diameter due to the confinement by the inner diameter of nanotube, though a few larger spheres (~120 nm) were seen across neighboring tube openings. While on the superhydrophobic areas (Fig. 16d), there was almost no CdS crystal. The high growth selectivity was also confirmed by the EDS analysis, revealing that CdS spheres easily nucleate and grow on the hydroxyl groups (–OH) terminated regions (Fig. 16e), but not on the –CF$_3$ terminated areas (Fig. 16f). Since the difference of the water contact angle between the superhydrophilic and superhydrophobic regions is larger than 150°, electrolyte solution is preferentially presented on the uniform superhydrophilic dots. No water droplets go to the neighboring superhydrophobic regions. Although a few CdS particles resulted from homogeneous precipitation attached onto superhydrophobic surface due to van der Waals interactions and gravity, they can be easily removed by ultrasonication. Therefore, a clear and well-defined CdS pattern in line with the dimensions of the superhydrophobic-superhydrophilic template has been obtained.

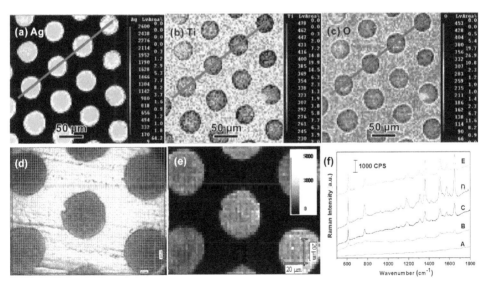

Fig. 15. Typical element distribution maps of Ag (a), Ti (b), O (c), and optical microscopy (d) and the corresponding R6G SERS mapping (e) of the patterned Ag NP films with an area of 140 × 100 μm^2 using the peak area at 614 cm^{-1} as the reference. (f) Raman spectra of the patterned Ag NP films from different charge density: curve A, bare subsrate; curve B, 0.5 C/cm^2; C, 1.0 C/cm^2; D, 2.0 C/cm^2; and E, 2.5 C/cm^2.

Fig. 16b shows the photocurrent spectra of the couple CdS/TiO$_2$ nanotube array electrode prepared under different electrodeposition times. It is apparent that the pure TiO$_2$ nanotube array samples have a photo-response wavelength lower than 400 nm due to its band-gap of 3.2 eV (curve a). The decoration of CdS nanospheres with a smaller energy band-gap (2.4 eV) can significantly extend the photo-response range from 380 nm to about 500 nm. Moreover, the CdS modified TiO$_2$ nanotube array electrodes can also greatly increase the photocurrent response under UV light, especially for the samples obtained under 2 min electrodeposition (curve b), which thus would be the optimal deposition time. This is attributed to the uniform dispersed CdS nanospheres with suitable size decorated onto the TiO$_2$ nanotubes. This allows for more efficient electron transfer and lower electron-hole recombination rate which leads to enhanced light harvesting at the directly grown CdS/TiO$_2$ heterojunctions. With the increase of time (curve c and d), more CdS particles with bigger size started to randomly distribute on top of TiO$_2$ nanotube arrays. Such composite nanostructures would weaken the light absorption of the uniform CdS/TiO$_2$ heterojunction underlayer, which has resulted in a lower photocurrent in both UV and visible light region.

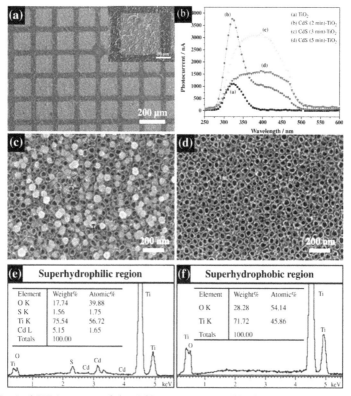

Fig. 16. (a) Typical SEM images of the CdS micropattern; (b) Photocurrent spectra of micropatterned CdS film on TiO$_2$ nanotube array electrode. (curve a-d): pure TiO$_2$; 2 min; 3 min; and 5 min. (c) Superhydrophilic region; (d) superhydrophobic region. EDX spectrum of the corresponding superhydrophilic (e) and superhydrophobic regions (f).

5. Summary and outlook

Extremely wetting micropattern (superhydrophilic/superhydrophobic) on TiO_2 nanostructure surface by using SAM technique and photocatalytic lithography has been studied intensely as it provides a cost effective template to construct well defined functional composited pattern. Numerous potential applications have also been proposed and investigated in biomedical, sensors and micro-nano devices. We believe that the photocatatlytic lithography patterning technique presented in this chapter should be general to create micro-scale wetting pattern on other semiconductor substrates and these developments will open the door for more widespread application of the wetting pattern in practical fields.

6. Acknowledgments

The authors thank the National Natural Science Foundation of China (grants 51072170, 21021002), the National Basic Research Program of China (grant 2007CB935603) and the National High Technology Research and Development Program of China (grant 2009AA03Z327), and the Environment and Water Industry Programme Office (EWI) under the National Research Foundation of Singapore (grant MEWR651/06/160) for their financial supports.

7. References

Balaur, E., Macak, J.M., Taveira, L. & Schmuki, P. (2005). Tailoring the wettability of TiO_2 nanotube layers. *Electrochem. Commun.*, Vol. 7, No. 10, 1066-1070, 1388-2481

Bearinger, J.P., Stone, G, Hiddessen, A.L., Dugan, L.C., Wu, L.G., Hailey, P., Conway, J.W., Kuenzler, T., Feller, L., Cerritelli, S. & Hubbell, J.A. (2009). Phototocatalytic lithography of poly(propylene sulfide) block copolymers: Toward high-throughput nanolithography for biomolecular arraying applications. *Langmuir*, Vol. 25, No. 2, 1238-1244, 0743-7463

Bhawalkar, S.P., Qian, J., Heiber, M.C. & Jia, L. (2010). Development of a colloidal lithography method for patterning nonplanar surfaces. *Langmuir*, Vol. 26, No. 22, 16662-16666, 0743-7463

Crawford, G.A. & Chawla, N. (2009). Porous hierarchical TiO_2 nanostructures: Processing and microstructure relationships. *Acta Mater.*, Vol. 57, No. 3, 854-867, 1359-6454

Csucs, G., Kunzler, T., Feldman, K., Robin, F. & Spencer, N.D. (2003). Microcontact printing of macromolecules with submicrometer resolution by means of polyolefin stamps. *Langmuir*, Vol. 19, No. 15, 6104-6109, 0743-7463

Cui, B. & Veres, T. (2007). Fabrication of metal nanoring array by nanoimprint lithography (NIL) and reactive ion etching. *Microelectron. Eng.*, Vol. 84, No. 5-8, 1544-1547, 0167-9317

Falconnet, D., Koenig, A., Assi, T. & Textor M. (2004). A combined photolithographic and molecular-assembly approach to produce functional micropatterns for applications in the biosciences. *Adv. Funct. Mater.*, Vol. 14, No. 8, 749-756, 1616-301X

Felidj, N., Aubard, J. & Levi, G. (1999). Discrete dipole approximation for ultraviolet-visible extinction spectra simulation of silver and gold colloids. *J. Chem. Phys.*, Vol. 111, No. 3, 1195-1208, 0021-9606

Gao, X.F. & Jiang, L. (2004). Water-repellent legs of water striders. *Nature,* Vol. 432, No. 7013, 36-36, 0028-0836

Gong, D.W., Grimes, C.A. & Varghese O.K. (2001). Titanium oxide nanotube arrays prepared by anodic oxidation. *J. Mater. Res.,* Vol. 16, No. 12, 3331-3334, 0884-2914

Haick, H. & Paz, Y. (2001). Remote photocatalytic activity as probed by measuring the degradation of self-assembled monolayers anchored near microdomains of titanium dioxide. *J. Phys. Chem. B,* Vol. 105, No. 15, 3045-3051, 1089-5647

Huang, L., Braunschweig, A.B., Shim, W., Qin, L.D., Lim, J.K., Hurst, S.J., Huo, F.W., Xue, C., Jong, J.W. & Mirkin, C.A. (2010). Matrix-assisted dip-pen nanolithography and polymer pen lithography. *Small,* Vol. 6, No. 10, 1077-1081, 1613-6810

Huang, Y.X., Lai, Y.K., Lin, L.X., Sun, L. & Lin, C.J. (2010). Electrochemical construction and biological performance of micropatterned CaP films. *Acta Phys. -Chim. Sin.,* Vol. 26, No. 8, 2057-2060, 1000-6818

Huang, Y.X., Sun, L., Xie, K.P., Lai, Y.K., Liu, B.J., Ren, B. & Lin, C.J. (2011). A SERS study of Ag nanoparticles electrodeposited on patterned TiO_2 nanotube films, *J. Raman Spectrosc.,* DOI: 10.1002/jrs.2830, 0377-0486, Vol. 42, No. 5, 986-991

Ichimura, K., Oh, S.K. & Nakagawa, M. (2000). Light-driven motion of liquids on a photoresponsive surface. *Science,* Vol. 288, No. 5471, 1624-1626, 0036-8075

Jeon, N.L., Clem, P.G., Payne, D.A. & Nuzzo, R.G. (1996). A monolayer-based lift-off process for patterning chemical vapor deposition copper thin films. *Langmuir,* Vol. 12, No. 22, 5350-5355, 0743-7463

Jiao, L.Y., Gao, H.J., Zhang, G.M., Xie, G.Y., Zhou, X., Zhang, Y.Y., Zhang, Y.Y., Gao, B., Luo, G., Wu, Z.Y., Zhu, T., Zhang, J., Liu, Z.F., Mu, S.C., Yang, H.F. & Gu, C.Z. (2005). Fabrication of metallic nanostructures by negative nanoimprint lithography. *Nanotechnology,* Vol. 16, No. 12, 2779-2784, 0957-4484

Katagiri, K., Ohno, K., Masuda, Y. & Koumoto, K. (2007). Growth behavior of TiO_2 particles via the liquid phase deposition process. *J. Ceram. Soc. Jpn.,* Vol. 115, No. 1348, 831-834, 1882-0743

Kobayashi, T., Shimizu, K., Kaizuma, Y. & Konishi, S. (2011). Formation of superhydrophobic/superhydrophilic patterns by combination of nanostructure-imprinted perfluoropolymer and nanostructured silicon oxide for biological droplet generation. *Appl. Phys. Lett.,* Vol. 98, No. 12, 123706, 0003-6951

Koumoto, K., Saito, N., Gao, Y.F., Masuda, Y. & Zhu, P.X. (2008). Nano/micro patterning of inorganic thin films. *Bull. Chem. Soc. Jpn.,* Vol. 81, No. 11, 1337-1376, 0009-2673

Kubo, W., Tatsuma, T., Fujishima, A. & Kobayashi, H. (2004). Mechanisms and resolution of photocatalytic lithography. *J. Phys. Chem. B,* Vol. 108, No. 9, 3005-3009, 1520-6106

Kumar, A., Biebuyck, H.A., Abbott, N.L. & Whitesides, G.M. (1992). The use of self-assembled monolayers and a selective etch to generate patterned gold features. *J. Am. Chem. Soc.,* Vol. 114, No. 23, 9188-9189, 0002-7863

Lafuma, A. & Quéré, D. (2003). Superhydrophobic states. *Nat. Mater.,* Vol. 2, No. 7, 457-460, 1476-1122

Li, L.Y., Yang, Y., Yang, G.L., Chen, X.M., Hsiao, B.S., Chu, B., Spanier, J.E. & Li, C.Y. (2006). Patterning polyethylene oligomers on carbon nanotubes using physical vapor deposition. *Nano Lett.,* Vol. 6, No. 5, 1007-1012, 1530-6984

Lai, Y.K., Sun, L., Zuo, J. & Lin, C.J. (2004). Electrochemical fabrication and formation mechanism of TiO_2 nanotube array on metallic titanium surface. *Acta Phys. -Chim. Sin.*, Vol. 20, No. 9, 1063-1066, 1000-6818

Lai, Y.K., Sun, L., Chen, Y.C., Zhuang, H.F., Lin, C.J. & Chin, J.W. (2006). Effects of the structure of TiO_2 nanotube array on Ti substrate on its photocatalytic activity. *J. Electrochem. Soc.*, Vol. 153, No. 7, D123-127, 0013-4651

Lai, Y.K., Lin, C.J., Wang, H., Huang, J.Y., Zhuang, H.F. & Sun, L. (2008). Superhydrophilic-superhydrophobic micropattern on TiO_2 nanotube films by photocatalytic lithography. *Electrochem. Commun.*, Vol. 10, No. 3, 387-391, 1388-2481

Lai, Y.K., Lin, C.J., Huang, J.Y., Zhuang, H.F., Sun, L. & Nguyen, T. (2008). Markedly controllable adhesion of superhydrophobic spongelike nanostructure TiO_2 films. *Langmuir*, Vol. 24, No. 8, 3867-3873, 0743-7463

Lai, Y.K., Gao, X.F., Zhuang, H.F., Huang, J.Y., Lin, C.J. & Jiang, L. (2009). Designing superhydrophobic porous nanostructures with tunable water adhesion. *Adv. Mater.*, Vol. 21, No. 37, 3799-3803, 0935-9648

Lai, Y.K., Huang, J.Y., Gong, J.J., Huang, Y.X., Wang, C.L., Chen, Z. & Lin, C.J. (2009). Superhydrophilic-superhydrophobic Template: A simple approach to micro- and nanostructure patterning of TiO_2 films. *J. Electrochem. Soc.*, Vol. 156, No. 11, D480-484, 0013-4651

Lai, Y.K., Zhuang, H.F., Sun, L., Chen, Z. & Lin, C.J. (2009). Self-organized TiO_2 nanotubes in mixed organic-inorganic electrolytes and their photoelectrochemical performance. *Electrochim. Acta*, Vol. 54, No. 26, 6536-6542, 0013-4686

Lai, Y.K., Lin, Z.Q., Huang, J.Y., Sun, L., Chen, Z. & Lin, C.J. (2010). Controllable construction of ZnO/TiO_2 patterning nanostructures by superhydrophilic/superhydrophobic templates. *New J. Chem.*, Vol. 34, No. 1, 44-51, 1144-0546

Lai, Y.K., Huang, J.Y., Zhang, H.F., Subramaniam, V.P., Tang, Y.X., Gong, D.G., Sundar, L., Sun, L., Chen, Z. & Lin, C.J. (2010). Nitrogen-doped TiO_2 nanotube array films with enhanced photocatalytic activity under various light sources. *J. Hazard. Mater.*, Vol. 184, No. 1-3, 855-863, 0304-3894

Lai, Y.K., Huang, Y.X., Huang, J.Y., Wang, H., Chen, Z. & Lin, C.J. (2010). Selective formation of ordered arrays of octacalcium phosphate ribbons on TiO_2 nanotube surface by template-assisted electrodeposition. *Colliods Surf. B*, Vol. 76, No. 1, 117-122, 0927-7765

Lai, Y.K., Lin, Z.Q., Chen, Z., Huang, J.Y. & Lin, C.J. (2010). Fabrication of patterned CdS/TiO_2 heterojunction by wettability template assisted electrodeposition. *Mater. Lett.*, Vol. 64, No. 11, 1309-1312, 0167-577X

Lee, J.P. & Sung, M.M. (2004). A new patterning method using photocatalytic lithography and selective atomic layer deposition. *J. Am. Chem. Soc.*, Vol. 126, No. 1, 28-29, 0002-7863

Lee, S.W., Oh, B.K., Sanedrin, R.G., Salaita, K., Fujigaya, T. & Mirkin, C.A. (2006). Biologically active protein nanoarrays generated using parallel dip-pen nanolithography. *Adv. Mater.*, Vol. 18, No. 9, 1133-1136, 0935-9648

Li, Z.W., Gu, Y.N., Wang, L., Ge, H.X., Wu, W., Xia, Q.F., Yuan, C.S., Chen, Y., Cui, B. & Williams, R.S. (2009). Hybrid nanoimprint-soft lithography with sub-15 nm resolution. *Nano Lett.*, Vol. 9, No. 6, 2306-2310, 1530-6984

Liu, S.H., Wang, W.C.M., Mannsfeld, S.C.B., Locklin, J., Erk, P., Gomez, M., Richter, F. & Bao, Z.N. (2007). Solution-assisted assembly of organic semiconducting single crystals on surfaces with patterned wettability. *Langmuir,* Vol. 23, No. 14, 7428-7432, 0743-7463

Liu, M.J., Zheng, Y.M., Zhai, J. & Jiang, L. (2010). Bioinspired super-antiwetting interfaces with special liquid-solid adhesion. *Accounts Chem. Res.,* Vol. 43, No. 3, 368-377, 0001-4842

Liu, X.J., Ye, Q.A., Yu, B., Liang, Y.M., Liu, W.M. & Zhou, F. (2010). Switching water droplet adhesion using responsive polymer brushes. *Langmuir,* Vol. 26, No. 14, 12377-12382, 0743-7463

Lu, Y., Liu, G.L. & Lee, L.P. (2005). High-density silver nanoparticle film with temperature-controllable interparticle spacing for a tunable surface enhanced Raman scattering substrate. *Nano Lett.,* Vol. 5, No. 1, 5-9, 1530-6984

Masuda, Y., Kinoshita, N., Sato, F. & Koumoto, K. (2006). Site-selective deposition and morphology control of UV- and visible-light-emitting ZnO crystals. *Cryst. Growth Des.,* Vol. 6, No. 1, 75-78, 1528-7483

Michel, R., Reviakine, I., Sutherland, D., Fokas, C., Csucs, G., Danuser, G., Spencer, N.D. & Textor, M. (2002). A novel approach to produce biologically relevant chemical patterns at the nanometer scale: Selective molecular assembly patterning combined with colloidal lithography. *Langmuir,* Vol. 18, No. 22, 8580-8586, 0743-7463

Nakata, K., Nishimoto, S., Yuda, Y., Ochiai, T., Murakami, T. & Fujishima, A. (2010). Rewritable superhydrophilic-superhydrophobic patterns on a sintered titanium dioxide substrate. *Langmuir,* Vol. 26, No. 14, 11628-11630, 0743-7463

Nishimoto, S., Sekine, H., Zhang, X.T., Liu, Z.Y., Nakata, K., Murakami, T., Koide, Y. & Fujishima, A. (2009). Assembly of self-assembled monolayer-coated Al_2O_3 on TiO_2 thin films for the fabrication of renewable superhydrophobic-superhydrophilic structures. *Langmuir,* Vol. 25, No. 13, 7226-7228, 0743-7463

Rausch, N. & Burte, E.P. (1993). Thin TiO_2 films prepared by low-pressure chemical vapor-deposition. *J. Electrochem. Soc.,* Vol. 140, No. 1, 145-149, 0013-4651

Rusponi, S., Costantini, G., de Mongeot, F.B., Boragno, C. & Valbusa, U. (1999). Patterning a surface on the nanometric scale by ion sputtering. *Appl. Phys. Lett.,* Vol. 75, No. 21, 3318-3320, 0003-6951

Shen, G.X., Chen, Y.C. & Lin, C.J. (2005). Corrosion protection of 316 L stainless steel by a TiO_2 nanoparticle coating prepared by sol-gel method. *Thin Solid Films,* Vol. 489, No. 1-2, 130-136, 0040-6090

Slocik, J.M., Beckel, E.R., Jiang, H., Enlow, J.O., Zabinski, J.S., Bunning, T.J. & Naik, R.R. (2006). Site-speciric patterning of biomolecules and quantum dots on functionalized surfaces generated by plasma-enhanced chemical vapor deposition. *Adv. Mater.,* Vol. 18, No. 16, 2095-2100, 0935-9648

Takeda, S., Suzuki, S., Odaka, H. & Hosono, H. (2001). Photocatalytic TiO_2 thin film deposited onto glass by DC magnetron sputtering. *Thin Solid Films,* Vol. 392, No. 2, 338-344, 0040-6090

Tatsuma, T., Kubo, W. & Fujishima, A. (2002). Patterning of solid surfaces by photocatalytic lithography based on the remote oxidation effect of TiO_2. *Langmuir,* Vol. 18, No. 25, 9632-9634, 0743-7463

Wang, H., Lin, C.J., Hu, R., Zhang, F. & Lin, L.W. (2008). A novel nano-micro structured octacalcium phosphate/protein composite coating on titanium by using an electrochemically induced deposition. *J. Biomed. Mater. Res. A*, Vol. 87A, No. 3, 698-705, 1549-3296

Wang, H., Duan, J.C. & Cheng, Q. (2011). Photocatalytically patterned TiO$_2$ arrays for on-plate selective enrichment of phosphopeptides and direct MALDI MS analysis. *Anal. Chem.*, Vol. 83, No. 5, 1624-1631, 0003-2700

Wang, R., Hashimoto, K., Fujishima, A., Chikuni, M., Kojima, E., Kitamura, A., Shimohigoshi, M. & Watanabe, T. (1997). Light-induced amphiphilic surfaces. *Nature*, Vol. 388, No. 6641, 431-432, 0028-0836

Wang, Y.L. & Lieberman, M. (2003). Growth of ultrasmooth octadecyltrichlorosilane self-assembled monolayers on SiO$_2$. *Langmuir*, Vol. 19, No. 4, 1159-1167, 0743-7463

West, J.B. (1999). The original presentation of Boyle's law. *J. Appl. Physiol.*, Vol. 87, No. 4, 1543-1545, 8750-7587

Xu, S. & Liu, G.Y. (1997). Nanometer-scale fabrication by simultaneous nanoshaving and molecular self-assembly. *Langmuir*, Vol. 13, No. 2, 127-129, 0743-7463

Yang, Y., Lai, Y.K., Zhang, Q.Q., Wu, K., Zhang, L.H., Lin, C.J. & Tang, P.F. (2010). A novel electrochemical strategy for improving blood compatibility of titanium-based biomaterials. *Colloids Surf. B*, Vol. 79, No. 1, 309-313, 0927-7765

Yasuda, K., Macak, J.M., Berger, S., Ghicov, A. & Schmuki, P. (2007). Mechanistic aspects of the self-organization process for oxide nanotube formation on valve metals. *J. Electrochem. Soc.*, Vol. 154, No. 9, C472-478, 0013-4651

Yoshimura, M. & Gallage, R. (2008). Direct patterning of nanostructured ceramics from solution - differences from conventional printing and lithographic methods. *J. Solid State Electrochem.*, Vol. 12, No. 7-8, 775-782, 1432-8488

Yun, H., Lin, C.J., Li, J. Wang, J.R. & Chen, H.B. Low-temperature hydrothermal formation of a net-like structured TiO$_2$ film and its performance of photogenerated cathode protection. *Appl. Surf. Sci.*, Vol. 255, No. 5, 2113-2117, 0169-4332

Zeng, J.B., Jia, H.Y., An, J., Han, X.X., Xu, W.Q., Zhao, B. & Ozaki, Y. (2008). Preparation and SERS study of triangular silver nanoparticle self-assembled films. *J. Raman Spectrosc.*, Vol. 39, No. 11, 1673-1678, 0377-0486

Zhang, C. & Kalyanaraman, R. (2004). In situ lateral patterning of thin films of various materials deposited by physical vapor deposition. *J. Mater. Res.*, Vol. 19, No. 2, 595-599, 0884-2914

Zhang, G.M., Zhang, J., Xie, G.Y., Liu, Z.F. & Shao H.B. (2006). Cicada wings: A stamp from nature for nanoimprint lithography. *Small*, Vol. 2, No. 12, 1440 1443, 1613 6810

Zhang, G.J., Tanii, T., Kanari, Y. & Ohdomari, I. (2007). Production of nanopatterns by a combination of electron beam lithography and a self-assembled monolayer for an antibody nanoarray. *J. Nanosci. Nanotechnol.*, Vol. 7, No. 2, 410-417, 1533-4880

Zhang, X.T., Jin, M., Liu, Z.Y., Tryk, D., Nishimoto, S., Murakami, T. & Fujishima, A. (2007). Superhydrophobic TiO$_2$ surfaces: Preparation, photocatalytic wettability conversion, and superhydrophobic-superhydrophilic patterning. *J. Phys. Chem. C*, Vol. 111, No. 39, 14521-14529, 1932-7447

Zhao, J.C., Wu, T.X., Wu, K.Q., Oikawa, K., Hidaka, H. & Serpone, N. (1998). Photoassisted degradation of dye pollutants. 3. Degradation of the cationic dye rhodamine B in aqueous anionic surfactant/TiO_2 dispersions under visible light irradiation: Evidence for the need of substrate adsorption on TiO_2 particles. *Environ. Sci. Technol.*, Vol. 32, No. 16, 2394-2400, 0013-936X

Zhuang, H.F., Lin, C.J., Lai, Y.K., Sun, L. & Li, J. (2007). Some critical structure factors of titanium oxide manotube array in its photocatalytic activity. *Environ. Sci. Technol.*, Vol. 41, No. 13, 4735-4740, 0013-936X

Permissions

The contributors of this book come from diverse backgrounds, making this book a truly international effort. This book will bring forth new frontiers with its revolutionizing research information and detailed analysis of the nascent developments around the world.

We would like to thank Bo Cui, for lending his expertise to make the book truly unique. He has played a crucial role in the development of this book. Without his invaluable contribution this book wouldn't have been possible. He has made vital efforts to compile up to date information on the varied aspects of this subject to make this book a valuable addition to the collection of many professionals and students.

This book was conceptualized with the vision of imparting up-to-date information and advanced data in this field. To ensure the same, a matchless editorial board was set up. Every individual on the board went through rigorous rounds of assessment to prove their worth. After which they invested a large part of their time researching and compiling the most relevant data for our readers. Conferences and sessions were held from time to time between the editorial board and the contributing authors to present the data in the most comprehensible form. The editorial team has worked tirelessly to provide valuable and valid information to help people across the globe.

Every chapter published in this book has been scrutinized by our experts. Their significance has been published with permission under the Creative Commons Attribution License or equivalent. extensively debated. The topics covered herein carry significant findings which will fuel the growth of the discipline. They may even be implemented as practical applications or may be referred to as a beginning point for another development. Chapters in this book were first published by InTech; hereby

The editorial board has been involved in producing this book since its inception. They have spent rigorous hours researching and exploring the diverse topics which have resulted in the successful publishing of this book. They have passed on their knowledge of decades through this book. To expedite this challenging task, the publisher supported the team at every step. A small team of assistant editors was also appointed to further simplify the editing procedure and attain best results for the readers.

Our editorial team has been hand-picked from every corner of the world. Their multi-ethnicity adds dynamic inputs to the discussions which result in innovative outcomes. These outcomes are then further discussed with the researchers and contributors who give their valuable feedback and opinion regarding the same. The feedback is then collaborated with the researches and they are edited in a comprehensive manner to aid the understanding of the subject.

Apart from the editorial board, the designing team has also invested a significant amount of their time in understanding the subject and creating the most relevant covers. They scrutinized every image to scout for the most suitable representation of the subject and create an appropriate cover for the book.

The publishing team has been involved in this book since its early stages. They were actively engaged in every process, be it collecting the data, connecting with the contributors or procuring relevant information. The team has been an ardent support to the editorial, designing and production team. Their endless efforts to recruit the best for this project, has resulted in the accomplishment of this book. They are a veteran in the field of academics and their pool of knowledge is as vast as their experience in printing. Their expertise and guidance has proved useful at every step. Their uncompromising quality standards have made this book an exceptional effort. Their encouragement from time to time has been an inspiration for everyone.

The publisher and the editorial board hope that this book will prove to be a valuable piece of knowledge for researchers, students, practitioners and scholars across the globe.

List of Contributors

Keita Sakai
Canon Inc., Japan

Sho Amano
University of Hyogo, Laboratory of Advanced Science and Technology for Industry (LAS-TI), Japan

J.P. Allain
Purdue University, United States of America

Christoph Ludwig and Steffen Meyer
Q-CELLS SE, Bitterfeld-Wolfen, Germany

Prasad Dasari, Jie Li, Jiangtao Hu and Nigel Smith
Nanometrics, USA

Oleg Kritsun
Globalfoundries, USA

SunHyung Lee and Katsuya Teshima
Faculty of Engineering, Shinshu University, Japan

Takahiro Ishizaki
National Institute of Advanced Industrial Science and Technology (AIST), Japan

Nagahiro Saito and Osamu Takai
EcoTopia Science Institute, Nagoya University, Japan

Yujun Song
Key Laboratory for Aerospace Materials and Performance (Ministry of Education), School of Materials Science and Engineering, Beihang University, Beijing, China

S. Sadegh Hassani and H. R. Aghabozorg
Research Institute of Petroleum Industry, Iran

Kyung-Hyun Choi, Khalid Rahman, Nauman Malik Muhammad, Arshad Khan, Ki-Rin Kwon, Yang-Hoi Doh and Hyung-Chan Kim
Jeju National University, Republic of Korea

Zhenyang Zhong, Tong Zhou, Yiwei Sun and Jie Lin
Fudan University/Department of Physics Shanghai, China

Hideo Kaiju
Research Institute for Electronic Science, Hokkaido University, Japan
PRESTO, Japan Science and Technology Agency, Japan

Kenji Kondo and Akira Ishibashi
Research Institute for Electronic Science, Hokkaido University, Japan

Yuekun Lai
State Key Laboratory for Physical Chemistry of Solid Surfaces and College of Chemistry and Chemical Engineering, Xiamen University, China
School of Materials Science and Engineering, Nanyang Technological University, Singapore

Changjian Lin
State Key Laboratory for Physical Chemistry of Solid Surfaces and College of Chemistry and Chemical Engineering, Xiamen University, China

Zhong Chen
School of Materials Science and Engineering, Nanyang Technological University, Singapore

Printed in the USA
CPSIA information can be obtained
at www.ICGtesting.com
JSHW011448221024
72173JS00004B/986